Springer Tracts in Modern Physics
Volume 140

Springer-Verlag Berlin Heidelberg GmbH

Springer Tracts in Modern Physics

Covering reviews with emphasis on the fields of Elementary Particle Physics, Solid-State Physics, Complex Systems, and Fundamental Astrophysics

Manuscripts for publication should be addressed to the editor mainly responsible for the field concerned:

Gerhard Höhler
Institut für Theoretische Teilchenphysik
Universität Karlsruhe
Postfach 6980
D-76128 Karlsruhe
Germany
Fax: +49 (7 21) 37 07 26
Phone: +49 (7 21) 6 08 33 75
Email: gerhard.hoehler@physik.uni-karlsruhe.de

Johann Kühn
Institut für Theoretische Teilchenphysik
Universität Karlsruhe
Postfach 6980
D-76128 Karlsruhe
Germany
Fax: +49 (7 21) 37 07 26
Phone: +49 (7 21) 6 08 33 72
Email: johann.kuehn@physik.uni-karlsruhe.de

Thomas Müller
IEKP
Fakultät für Physik
Universität Karlsruhe
Postfach 6980
D-76128 Karlsruhe
Germany
Fax:+49 (7 21) 6 07 26 21
Phone: +49 (7 21) 6 08 35 24
Email: mullerth@vxcern.cern.ch

Roberto Peccei
Department of Physics
University of California, Los Angeles
405 Hilgard Avenue
Los Angeles, California 90024-1547
USA
Fax: +1 310 825 9368
Phone: +1 310 825 1042
Email: robertop@college.ucla.edu

Frank Steiner
Abteilung für Theoretische Physik
Universität Ulm
Albert-Einstein-Allee 11
D-89069 Ulm
Germany
Fax: +49 (7 31) 5 02 29 24
Phone: +49 (7 31) 5 02 29 10
Email: steiner@physik.uni-ulm.de

Joachim Trümper
Max-Planck-Institut
für Extraterrestrische Physik
Postfach 1603
D-85740 Garching
Germany
Fax: +49 (89) 32 99 35 69
Phone: +49 (89) 32 99 35 59
Email: jtrumper@mpe-garching.mpg.de

Peter Wölfle
Institut für Theorie
der Kondensierten Materie
Universität Karlsruhe
Postfach 69 80
D-76128 Karlsruhe
Germany
Fax: +49 (7 21) 69 81 50
Phone: +49 (7 21) 6 08 35 90/33 67
Email: woelfle@tkm.physik.uni-karlsruhe.de

James A. Crittenden

Exclusive Production of Neutral Vector Mesons at the Electron–Proton Collider HERA

With 34 Figures

 Springer

Dr. James A. Crittenden

Universität Bonn
Physikalisches Institut
Nußallee 12
D-53115 Bonn, Germany
E-mail: crittenden@physik.uni-bonn.de

Library of Congress Cataloging-in-Publication Data

Crittenden, James Arthur. Exclusive production of neutral vector mesons at the electron–proton collider HERA/ James A. Crittenden. p. cm. – (Springer tracts in modern physics; vol. 140) Includes bibliographical references and index.
ISBN 978-3-662-14809-9   ISBN 978-3-540-69251-5 (eBook)
DOI 10.1007/978-3-540-69251-5
I. Title. II. Series: Springer tracts in modern physics; 140. QC1.S797 vol. 140 539 s–dc21 [539.7'2162] 97-28373 CIP

Physics and Astronomy Classification Scheme (PACS):
14.70.Bh, 12.38.-t, 12.38.Qk, 13.60.-r, 13.85.Hd, 13.87,-a

ISBN 978-3-662-14809-9

© Springer-Verlag Berlin Heidelberg 1997
Originally published by Springer-Verlag Berlin Heidelberg New York in 1997
Softcover reprint of the hardcover 1st edition 1997

Typesetting: Camera-ready copy by the author using a Springer T_EX macro-package
Cover design: *design & production* GmbH, Heidelberg
SPIN: 10633596   56/3144-5 4 3 2 1 0 – Printed on acid-free paper

Preface

The first five years of operation of the multi-purpose experiments ZEUS and H1 at the electron–proton storage ring facility HERA have opened a new era in the study of vector-meson production in photon–proton interactions. Of particular interest are the investigations of *exclusive* vector-meson production, since such completely reconstructed final states are rare in experiments at high-energy colliders. The observed diffractive nature of these reactions identifies them as contributing to the new field of hard diffraction, a topic which has drawn much interest in recent years as a first opportunity for testing the extension of perturbative quantum chromodynamical calculations to diffractive processes. Since soft diffractive processes are studied in these HERA experiments as well, the transition between the two domains has now become experimentally accessible and can be studied as a function of the vector-meson mass, photon virtuality, momentum transferred to the proton, and center-of-mass photon–proton energy.

The high energies available at the unique accelerator complex HERA allow investigations in hitherto unexplored kinematic regions, providing answers to long-standing questions concerning the energy dependence of the ρ^0, ω, ϕ, and J/ψ production cross sections. The excellent angular acceptance of the central detectors, combined with that of specialized tagging detectors at small production angles, has permitted measurements of elastic and inelastic production processes for both quasi-real photons and those of virtuality far exceeding the squared mass of the vector meson. This review provides a quantitative picture of the present status of these studies, compares them to the extensive measurements in this field at lower energies, and summarizes topical developments in theoretical work motivated by the new data.

This work was immeasurably assisted by many pleasant discussions with my colleagues at the Deutsches Elektronen-Synchrotron laboratory. The final version bears the influence of constructive criticism from M. Arneodo, S. Bhadra, E. Gallo, U. Katz, R. Klanner, P. Marage, A.S. Proskuryakov, and A. Quadt. The difficult task of coordinating the efforts in the ZEUS diffractive physics working group was accomplished by M. Arneodo, S. Bhadra, E. Gallo, and G. Iacobucci. I extend thanks to T. Doeker, T. Monteiro, and Q. Zhu for explanations concerning the data from the ZEUS beam-pipe calorimeter. I found my discussions of the helicity analyses with H. Beier,

K. Piotrzkowski, A.S. Proskuryakov, D. Westphal, and G. Wolf particularly enlightening. A. Levy was of great assistance on the subject of photon–proton total cross sections. A. Quadt provided Fig. 1.4. Discussions with B. Burow on the vagaries of virtual photon flux definitions served as a calming influence.

I am further grateful to S. Brodsky, B. Kopeliovich, and L. Frankfurt for answering questions pertaining to the description of vector-meson production in the framework of perturbative QCD calculations.

The research reported in this review was made possible by the heroic efforts of the HERA accelerator physicists and the constant support provided to the HERA experiments by the DESY directorate.

This work was performed in the context of the research group led by Professor E. Hilger at the Physics Institute of the University of Bonn. I would like to acknowledge his unflagging encouragement and helpful commentary on several drafts of this review.

The research activities of the ZEUS group in Bonn are supported by the German Federal Ministry for Education and Science, Research and Technology (BMBF).

Bonn, July 1997   *James A. Crittenden*

Contents

1. Introduction

Experimentation at the accelerator complex HERA (Hadron-Elektron-Ring-Anlage) operating at the German research laboratory DESY (Deutsches Elektronen-Synchrotron) during the past five years has revealed an array of nature's secrets to perusal by the curiosity-driven human intellect. The opening of hitherto unexplored kinematic regions at high interaction energies is resulting in qualitative progress in our understanding of hadronic interactions. A variety of questions, both old and new, are being stimulated by the access to this information on the elementary particles and their interactions. Decades-old conceptual quandaries are once again being addressed as we witness renewed studies of electron–proton interactions. The studies at HERA have fulfilled the expectations of unprecedented breadth in the accessible fields of research. The classic studies of deep inelastic scattering and the structure of the proton have been extended to a kinematic domain in which our microscopes now resolve objects orders of magnitude smaller than the proton itself. The hadronic nature of the photon enjoys a rejuvenated interest stimulated not only by the recently obtained experimental information on photon–hadron interactions, but also by exciting developments in theoretical approaches to understanding these interactions. For the first time, the energy dependence of the hadronic coupling of the photon can be studied over a broad range of precisely measured photon virtualities. In particular, data samples of high statistical precision have now become available for study of the production of vector mesons in both exclusive and inclusive reactions. These vector mesons are objects both familiar and arcane: familiar due to extensive studies of their properties in the 1960s during an era of discovery in particle physics, as accelerators progressively achieved the energies necessary to exceed the production energy thresholds, and arcane because they carry the quantum numbers of the photon, resulting in quantum-mechanical mixing between the mesonic and photonic versions of this mysterious form of matter/energy, and yet represent the simplest form of quark bound states. Indeed, studies of deep inelastic electron–proton scattering stimulated the early development of the theory of quantum chromodynamics. This youngest addition to our field theoretical descriptions of the forces in nature remains after twenty-five years our most convincing conceptual framework for understanding the interactions between the objects we call quarks: objects which have never been observed as free particles, yet interact as such at high energy. The vector mesons ρ^0, ω, ϕ, J/ψ, and Υ represent the quark/anti-quark bound states corresponding to the generational hierarchy employed in the Standard Model. In this review we investigate the recent experimental and phenomenological developments in the understanding of vector-meson production in photon–proton interactions, profiting from the broad kinematic domain accessible at HERA, where a strict delineation between "hard" and "soft" interactions, motivated in part by experimental constraints in prior studies, is no longer sufficient, or even useful.

The content of this review must be considered as an intermediate step in the development of this field of study for the following two reasons. First, the vigorous pursuit of improved experimentation at HERA presently underway permits the expectation of more than an order-of-magnitude increase in statistical precision over the coming two years, an improvement certain to open qualitatively different avenues of investigation, as well as to sharpen the experimental results presented here. Second, the theoretical research addressing the above-mentioned distinction between hard and soft processes is still in its early stages and we can certainly allow ourselves the hope that major progress will be achieved in the coming years. However, the successful operation of the HERA research facilities during these past five years has clearly delineated the range of physics issues involved in the field of vector-meson production in photon–proton interactions, and it is time to take stock of the situation and prepare for the future. Such is the intent of the present work.

We begin by placing the HERA project in historical perspective, briefly describing in Sects. 1.1 and 1.2 the developments which led to the commissioning of the accelerator complex in 1992 and its peformance during the first five years of operation. Thereafter follow descriptions in Sect. 1.3 of the two multi-purpose detectors built and operated by the H1 and ZEUS collaborations. In Sect. 1.4 we provide an overview of the physics topics addressed at HERA to date, beginning with a review of the kinematics of electron–proton interactions. After a short synopsis in Sect. 1.5 of vector-meson production studies since the 1960s, we turn to a review of contemporary topical theoretical considerations in Chap. 2. Section 3 begins the discussion of experimental results with a description of the total photon–proton cross section measurements in Sect. 3.1 before turning to the production of single neutral vector mesons in Sect. 3.2. A discussion of general experimental considerations is followed by a detailed description of resolutions, acceptances, and backgrounds associated with each of the decay channels for which results have been published by the two collaborations. Observations on the energy and momentum transfer dependences of the production cross sections and a synopsis of results on production ratios are followed by a description of the helicity analyses leading to determinations of the ratio of longitudinal to transverse photoabsorption cross sections as a function of photon virtuality. This section on experimental results from HERA ends with brief descriptions of the measurements of vector-meson production accompanied by dissociation of the proton. Chapter 4 summarizes the foreseeable improvements in present measurements, discussing luminosity upgrade plans, additional detector components, prospects for beam polarization, and those for nuclear beams. The report itself is summarized in Chap. 5 and some general concluding remarks are offered in Chap. 6. This review covers the theoretical and experimental work published prior to the end of 1996. We use natural units ($\hbar = c = 1$).

1.1 Historical Context

A strong interest in the study of electron–proton interactions at the high energies provided by colliding beam accelerators followed the successes of the deep inelastic scattering experiments at the Stanford Linear Accelerator (SLAC) in the late 1960s. Both the PEP (Positron–Electron–Proton) accelerator facility at SLAC and its counterpart PETRA (Proton–Electron Tandem Ring Accelerator) at DESY were originally proposed as electron–proton colliding beam facilities [1]. However, the astonishing discoveries at the e^+e^- facility SPEAR at SLAC, beginning with that of the J/ψ vector meson in November, 1974, and continuing with the investigations of charmed mesons and of the τ lepton, turned the tide in favor of electron–positron colliding facilities, leading to the exhaustive studies of e^+e^- interactions at the PEP and PETRA accelerators throughout the mid-1980s and continuing today at the LEP accelerator at CERN (European Laboratory for Particle Physics). Among the prime motivating factors for the effort put into this research was the hope that the higher-mass quarks and their spectroscopy could be studied as thoroughly at these machines as were charmed quarks at the SPEAR facility. However, both the bottom and top quarks have since been discovered via their production in hadronic interactions. The electron–positron colliders CESAR at Cornell University and DORIS at DESY have indeed proved essential for the spectroscopy of the charm and bottom quarks and LEP continues today to add to the wealth of information on this rich topic. Concurrently, the continued study of deep inelastic scattering with neutrino and muon beams at the Fermi National Accelerator Laboratory (FNAL) in Chicago and CERN in Geneva were producing impressive contributions to the understanding of elementary particles and their interactions. The discovery of weak neutral currents in 1973 at CERN provided an important boost to the credibility of the electroweak Standard Model. Precision measurements of scaling violations in the interactions of muons and neutrinos with nucleons at both CERN and FNAL provided quantitative information necessary to the progress in the new theory of strong interactions, quantum chromodynamics. Thus the motivation to continue the early studies in electron–proton interactions remained strong, especially since advances in accelerator technology for both proton and electron beams made accessible kinematic regions in which the Standard Model predicted that the so-called "weak" coupling should be comparable to the electromagnetic coupling, far exceeding the energy range accessible in fixed-target applications. One crucial question remained: Could a solution be found for the hitherto unaddressed difficulties associated with focusing beams of vastly different characteristics to the minute transverse size necessary to obtain high interaction rates?

By the early 1980s the superconducting magnet development studies at FNAL had made clear the feasibility of 800 GeV proton accelerators, and this work was being applied to the construction of a symmetric p$\bar{\text{p}}$ storage ring accelerator complex. A proposal to add a 30 GeV electron ring to this facility [2] ended up being rejected in favor of FNAL's p$\bar{\text{p}}$ and fixed-target programs.

In 1979 the European Committee for Future Accelerators published a feasibility study for a high-energy electron–proton accelerator facility [3]. On the basis of this and further design studies at DESY, a proposal for a colliding beam facility featuring 30 GeV electrons and 800 GeV protons was submitted to German funding agencies in 1983 [4]. This proposal was approved in 1984, contingent on substantial financial and technical contributions from abroad. The success of the resulting international collaboration led to the coining of the term "HERA model" as a paradigm for future particle accelerator projects.

By the end of 1987 the civil construction work on the HERA tunnel and experimental halls had been completed, and a huge liquid helium plant for the superconducting proton ring magnets was operational. The electron ring was commissioned in mid-1988, as the industrial construction of the proton ring magnets was getting underway. The last of these magnets was installed in the tunnel in September, 1990. The first electron–proton collisions were achieved in October, 1991, and after the two large multi-purpose detectors ZEUS and H1 were commissioned during the spring of 1992, the first physics run of the machine took place in June.

1.2 The Performance of the HERA Accelerators

The rapid improvement in HERA performance [5] during its first five years of operation is described by the integrated luminosities per year shown in Table 1.1. Studies of limitations in the electron beam lifetime in 1992 and 1993 concluded that space charge and recombination effects involving positively charged ions in the beam pipe were to blame. Indeed the switch to positrons in early 1994 led to an immediate improvement in the beam lifetime and hence increased integrated luminosity. As a result, most of the results we discuss in this review were obtained with a positron beam, though we often use the term "electron" generically in referring to the scattered lepton. Further running with an electron beam is presently planned following extensive improvements to the electron ring vacuum system prior to the 1998 running period.

Table 1.1. Yearly integrated luminosities during the first five years of HERA operation

Year	Energy (GeV)		Integrated Luminosity (pb^{-1})	
	e	p	e^-p	e^+p
1992	26.7	820	0.05	–
1993	26.7	820	1	–
1994	27.5	820	1	5
1995	27.5	820	–	12
1996	27.5	820	–	17

1.3 The Two Multi-Purpose Detectors H1 and ZEUS

The technical specifications of the HERA accelerators posed unique problems not only for the machine physicists, but also for those designing the detectors. The bunch crossing rate of 10 MHz was a factor of forty greater than that of the CERN p$\bar{\text{p}}$ collider, resulting in new problems for data acquisition and triggering. The Lorentz boost of the center-of-mass reference frame in that of the laboratory necessitated the design of large-acceptance detectors more highly segmented and deeper in the proton flight direction. Sensitivity to interaction rates of a few events per day (for example, charged-current deep inelastic scattering) was required in the presence of background rates of 100 kHz from the interactions of high-energy protons with residual gas in the beam pipe. The calorimeters had to be designed for the measurements of hadronic jets of several hundred GeV as well as for those of muonic energy depositions of a few hundred MeV. In the following we will discuss briefly how each of the two multi-purpose detectors addresses these common problems with complementary methods, emphasizing the detector components necessary to the triggering and reconstruction of events with single neutral vector mesons in the final state. We will refer to a right-handed coordinate system in which the Z axis points in the proton flight direction, called the forward direction, and the X axis points toward the center of the HERA ring, with the origin at the nominal interaction point.

1.3.1 The H1 Detector

The design of the 2800 ton H1 detector [6], shown in Fig. 1.1, emphasizes charged particle tracking in the central region along with full calorimetric coverage and high resolution in the measurement of electromagnetic energy depositions. The primary components of the tracking system are two coaxial, cylindrical, jet-type drift chambers [7] covering the polar angle region between 20° and 160°. Position resolutions per hit of better than 200 µm in the radial direction and 2 cm in the longitudinal direction have been attained. Further inner drift chambers and two proportional chambers serve to measure the longitudinal track coordinates and to provide triggering information.

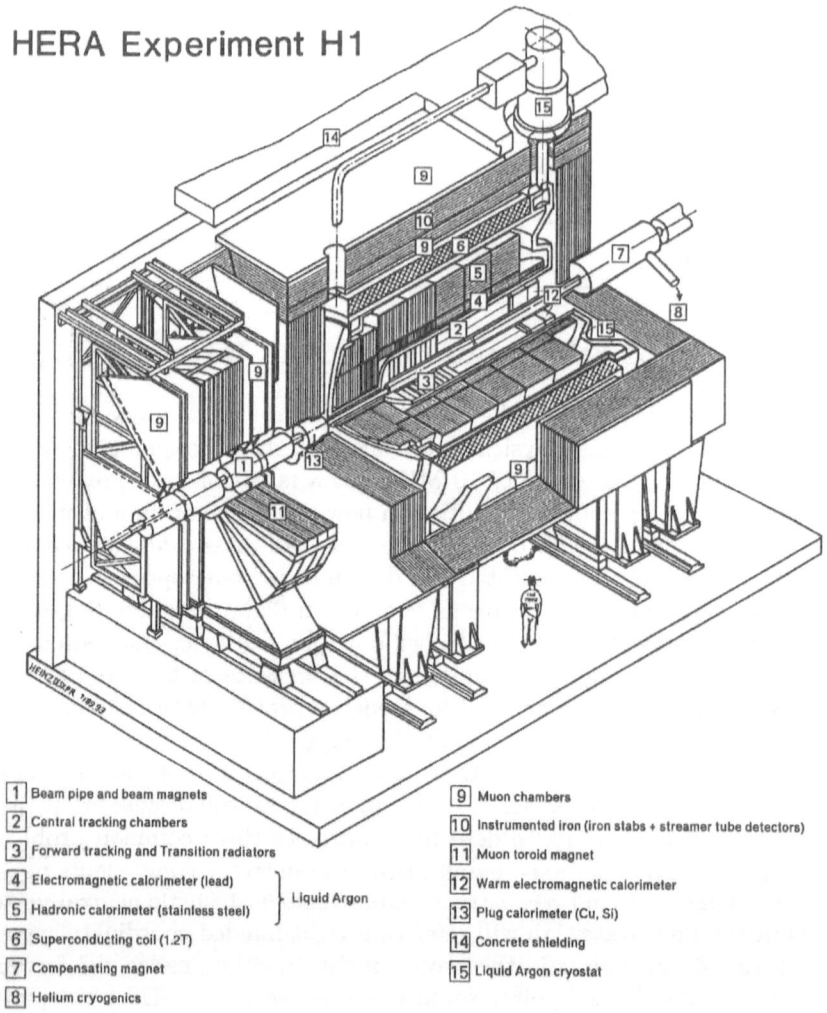

HERA Experiment H1

1 Beam pipe and beam magnets	9 Muon chambers
2 Central tracking chambers	10 Instrumented iron (iron slabs + streamer tube detectors)
3 Forward tracking and Transition radiators	11 Muon toroid magnet
4 Electromagnetic calorimeter (lead)	12 Warm electromagnetic calorimeter
5 Hadronic calorimeter (stainless steel)	13 Plug calorimeter (Cu, Si)
6 Superconducting coil (1.2T)	14 Concrete shielding
7 Compensating magnet	15 Liquid Argon cryostat
8 Helium cryogenics	

Liquid Argon

Fig. 1.1. The H1 detector

The tracking system achieves a momentum resolution in the coordinate transverse to the 1.2 T solenoidal field of $\sigma(p_{\mathrm{T}})/p_{\mathrm{T}} = 0.01\,p_{\mathrm{T}} \oplus 0.015$, where p_{T} is expressed in units of GeV. Specific ionization measurements provide information used for particle identification. The magnetic field is produced by an impressive 5-m-long superconducting solenoid of sufficient diameter (5.8 m) to enclose the entire calorimeter, storing 100 MJ of energy during operation. The H1 calorimeter [8] consists of a lead/liquid-argon shower-sampling sandwich section designed to measure electromagnetic energy depositions, followed by hadronic sections employing steel plates as absorbers. The depths of the hadronic sections range from 6 interaction lengths (λ_I) in the forward direction to 4 λ_I toward the rear. The angular coverage of the calorimeter extends from 4° to 155° in polar angle. The transverse segmentation of the electromagnetic calorimeter is 4×4 cm^2, taking advantage of the flexibility in electrode design of the liquid-argon calorimetric technique. The electromagnetic calorimeter is further segmented in the direction of longitudinal shower development into 3 (4) sections in the barrel (forward) regions. The transverse segmentation of the hadronic sections is 4×4 cm^2 and there are 4 (6) longitudinal sections in the barrel (forward) region. This fine segmentation results in 31 000 electromagnetic and 14 000 hadronic readout channels. The electromagnetic calorimeter achieves an energy resolution of $\sigma(E)/E = 0.12/\sqrt{E(\text{GeV})} \oplus 0.01$. The high degree of segmentation permits an ingenious reconstruction algorithm for hadronic energy depositions which reweights the electromagnetic and hadronic components of the shower so as to achieve the desired linearity and resolution characteristics despite sampling fractions which differ by 30% for the two components. The hadronic energy resolution thus obtained is $\sigma(E)/E = 0.55/\sqrt{E(\text{GeV})} \oplus 0.01$. During the first three years of operation, the rear region between polar angles of 150° and 175° was covered by a lead/scintillator-plate calorimeter with a resolution of $\sigma(E)/E = 0.10/\sqrt{E(\text{GeV})} \oplus 0.01$. In 1995 this calorimeter was replaced with a lead/scintillating-fiber calorimeter of transverse cell size 4×4 cm^2, energy resolution $\sigma(E)/E = 0.07/\sqrt{E(\text{GeV})} \oplus 0.01$, and time resolution better than 1 ns, which also included a hadronic compartment 1 λ_I thick. Muon identification is provided by limited-streamer tube instrumentation in the magnetic field return yoke outside the calorimeter cryostat, comprising 16 layers with a resolution of 1 cm per layer.

Our discussion of exclusive vector-meson production will necessitate the distinction of event classes with and without dissociation of the proton. The H1 measurements make use of two detectors in the extreme forward direction to distinguish experimentally these two event types: a toroidal muon spectrometer covering the polar angular region between 3° and 17° and a system of scintillators 24 m from the interaction point which cover the angular region from 0.06° to 0.25°. These devices provide sensitivity to secondary particles produced in the beam-pipe wall and other beam-line elements by interactions of the dissociated-proton system.

The H1 triggering technique relies on track reconstruction in the proportional chamber and drift chamber systems, which are efficient for tracks with transverse momenta exceeding a value of 0.5 GeV. Final-state electrons can be identified in the trigger system via depositions in the electromagnetic calorimeter for energies greater than 800 MeV and also via depositions in the small-angle luminosity monitor electron detector. Muons are identified at the trigger level via hits in the central muon system.

1.3.2 The ZEUS Detector

The design of the ZEUS detector [9], a schematic diagram of which is shown in Fig. 1.2, profited from innovative long-term test beam programs at CERN and FNAL which led to substantial progress in the understanding of hadron calorimetry [10]. In particular, the ability to design a uranium sampling calorimeter with equal sampling fractions for electromagnetic and hadronic shower

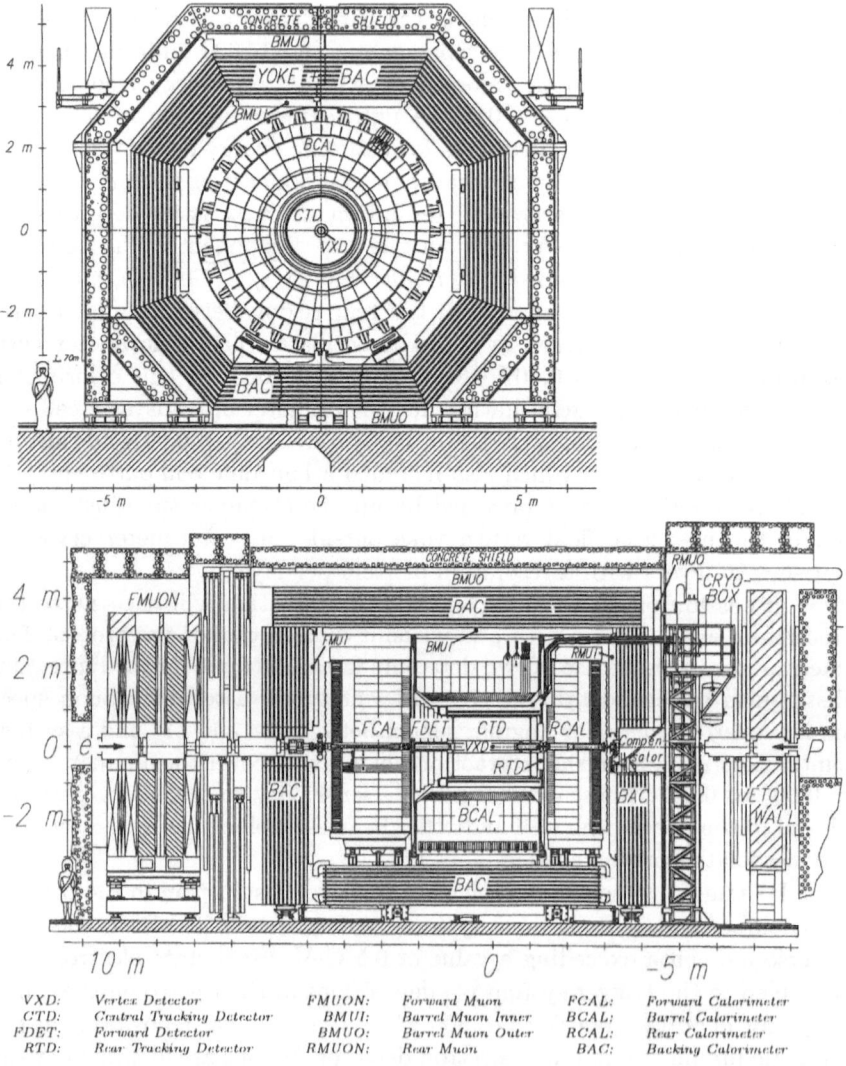

VXD:	Vertex Detector	FMUON:	Forward Muon	FCAL:	Forward Calorimeter
CTD:	Central Tracking Detector	BMUI:	Barrel Muon Inner	BCAL:	Barrel Calorimeter
FDET:	Forward Detector	BMUO:	Barrel Muon Outer	RCAL:	Rear Calorimeter
RTD:	Rear Tracking Detector	RMUON:	Rear Muon	BAC:	Backing Calorimeter

Fig. 1.2. The ZEUS detector

components (termed a "compensating" calorimeter [11]) permitted unprecedented specifications in hadronic energy resolution and linearity at the trigger level as well as for the offline reconstruction of hadron energies. This leads to one of the remarkable contrasts between these two HERA detectors: the ZEUS trigger algorithm is primarily calorimeter-based, while that of H1 emphasizes tracking algorithms for reconstruction of the interaction vertex. The 700-ton ZEUS calorimeter [12] consists of forward, central barrel, and rear sections, covering the polar angle ranges $2° - 40°$, $37° - 129°$, and $128° - 177°$. Each is made up of layers of 2.6 mm SCSN-38 scintillator and 3.3 mm stainless-steel-clad depleted-uranium plates. Such a layer corresponds to 1.0 radiation length (X_0) and 0.04 interaction lengths. This uniformity in structure throughout the entire calorimeter and the distributed nature of the natural radioactivity of the uranium used to set the gains of the phototubes are decisive factors in the stability and precision of its calibration [13]. The choice of active and passive thicknesses results in a sampling fraction of 4% for electromagnetic and hadronic shower components (hence compensation) and 7% for minimum-ionizing particles. The compensation results in a hadronic energy resolution of $\sigma(E)/E = 0.35/\sqrt{E(\text{GeV})} \oplus 0.02$. The resolution for electromagnetic showers is $\sigma(E)/E = 0.18/\sqrt{E(\text{GeV})} \oplus 0.01$.

The wavelength-shifting optical readout components subdivide the calorimeters longitudinally into electromagnetic and hadronic sections. The electromagnetic sections are 25 X_0 (1.1 λ_I) deep. The forward (barrel) hadronic sections are further segmented longitudinally into two sections, each of 3.0 λ_I (2.0 λ_I) depth. The rear calorimeter has a single hadronic section 3.0 λ_I deep. The optical components subdivide the forward and rear calorimeters into towers of transverse dimensions 20×20 cm^2. The electromagnetic sections are further divided vertically into four (two) cells in the forward (rear) calorimeter. Thus the transverse segmentation of the electromagnetic cells is 20×5 cm^2 in the forward region and 20×10 cm^2 in the rear. The front face of the forward (rear) calorimeter is 2.2 m (1.5 m) distant from the nominal electron–proton interaction point. The barrel calorimeter consists of 32 wedge-shaped modules situated 1.2 m distant from the beam axis. Its electromagnetic sections are divided by the optical readout components into 53 cells per module, each of interior transverse dimensions 24×5 cm^2, projective in polar angle. The 14 hadronic towers in each module measure 24×28 cm^2 in the interior transverse dimensions. A 2.5° rotation of all the barrel calorimeter modules relative to the radial direction ensures that the intermodular regions containing the wavelength shifters do not point at the beam axis.

The rear calorimeter is further instrumented with a layer of 3×3 cm^2 silicon diodes located at a depth of 3.3 X_0 in the electromagnetic section, an element essential to the reconstruction of low-energy neutral pions from ω decays.

The ZEUS solenoidal coil of diameter 1.9 m and length 2.6 m provides a 1.43 T magnetic field for the charged-particle tracking volume. The tracking system consists of a central jet-type wire chamber [14] covering the polar angular region from 15° to 164°, a forward planar tracking detector from 8° to 28°, and a second planar tracking chamber in the backward direction, covering the region from 158° to 170°. The momentum resolution attained is $\sigma(p_T)/p_T = 0.005\,p_T \oplus 0.015$ and a track is extrapolated to the calorimeter face with a transverse resolution of about 3 mm. Ionization measurements from the central tracking chamber also serve to identify electron–positron pairs from J/ψ decays.

The coil represents about 1 X_0 of passive absorber between the barrel calorimeter and the interaction region, similar to the amount of structural material in front of the forward and rear calorimeters. A small-angle rear tracking device constructed of 1-cm-wide scintillator strips and a forward and rear scintillator-tile device with 20×20 cm^2 segmentation serve as shower presampling detectors, improving the energy measurement for the scattered electron or positron [15].

The muon system [16] is constructed of limited-streamer tubes inside and outside the magnetic return yoke, covering the region in polar angle from 34° to 171°. Hits in the inner chambers provide muon triggers for J/ψ decays.

The shaping, sampling, and pipelining algorithm of the readout, developed for the calorimeter [17] and used in modified form for the silicon and presampler systems, permits the reconstruction of shower times with respect to the bunch crossings with a resolution of better than 1 ns, providing essential rejection against upstream beam–gas interactions, as well as allowing 5 μs for the calculations of the calorimeter trigger processor [18]. For the triggering of the simple final states associated with vector-meson decays in the exclusive production channels we discuss in this review a trigger threshold of 0.5 GeV was used in the calorimeter, along with track candidates in the central drift chamber, or a threshold of 0.5 GeV with hits in the muon chambers.

In 1994 the Leading Proton Spectrometer (LPS) [19, 20] began operation. This large-scale system of six stations of silicon-strip detectors located between 24 and 90 m from the interaction point employs "Roman pot" technology to place active detector planes within a few millimeters of the proton beam. Beamline magnets provide a resolving power of 0.4% in the longitudinal coordinate of proton momentum and 5 MeV in the momentum component transverse to the nominal proton beam axis. The beam transverse momentum distribution has r.m.s. values of 40 MeV in the horizontal plane and 90 MeV in the vertical plane, which dominate the LPS resolution. The acceptance in longitudinal momentum extends from 40% to 100% of the proton beam momentum. When used to identify exclusive vector-meson production, the LPS reduces contamination by processes in which the proton dissociates into a multiparticle final state to under 0.5%, compared to 10–20% achieved with the other detectors in the forward region. This measurement of the final-state proton also provides a direct measurement of the momentum transfer to the proton, a measurement more accurate than can be derived from the other detectors.

The Proton Remnant Tagger (PRT) consists of a set of seven pairs of scintillators situated 5.2 m from the interaction point in the forward direction. They cover polar angles between 6 and 26 mrad and serve to discriminate between exclusive vector-meson production processes and those in which the proton dissociates into a multiparticle final state producing depositions in the PRT.

Prior to the 1995 data-taking period the ZEUS collaboration installed a small tungsten/scintillator electromagnetic sampling calorimeter around the beam pipe in the rear direction 3 m from the nominal interaction point. Covering the region of small electron scattering angle between 17 and 35 mrad, this beam-pipe calorimeter (BPC) allows measurements for values of photon virtuality extending from 0.1 to 0.9 GeV^2. At the time of writing, analysis results are becoming available and promise to play an important role in understanding the transition region between photoproduction processes and deep inelastic scattering.

1.4 Investigations of Photon–Proton Interactions at HERA

This section provides a brief general overview of the issues in photon–proton interactions addressed by investigations at HERA. After an introduction to the notation and a discussion of the accessible ranges in the various kinematic variables, three broad categories of physics topic are treated. These categories will be seen to have some overlap and serve primarily as a rhetorical tool of organization rather than as indicators of physics-based distinction. The choice of these categories is furthermore conditional on the present-day state of HERA operation. As the instantaneous luminosity increases and data with other beam energies become available, this categorization of physics topics studied at HERA will require modification.

1.4.1 Kinematics

Much of the notation and choice of variables in this section is motivated by the investigations of deep inelastic electron–nucleon scattering performed at SLAC in the late 1960s which led to the discovery of Bjorken scaling [21] and the further development of the infinite momentum frame formalism [22, 23]. For a detailed general treatment of particle kinematics, extending beyond that of deep inelastic scattering, we refer to the book by Byckling and Kajantie [24].

We consider the neutral-current process $e^{\pm} + p \rightarrow e^{\pm} + X$, and choose the notations k, k', and P for the four-momenta of the initial-state electron, the final-state electron, and the initial-state proton respectively. The charged-current process with its final-state neutrino has also been measured at HERA; we restrict ourselves in this discussion to photon exchange, neglecting the contributions from exchange of the heavy intermediate vector bosons $W^{\pm}$ and Z^0. These contributions become important at virtualities comparable to squared vector-boson masses, whereas the present-day event statistics for the processes we consider in this review are limited to virtualities less than about 30 GeV^2. The Mandelstam variable s for the two-body scatter represents the square of the electron–proton center-of-mass energy

$$s \;\equiv\; (\boldsymbol{k} + \boldsymbol{P})^2 \;=\; m_{\mathrm{e}}^2 + m_{\mathrm{p}}^2 + 4\, E_{\mathrm{e}}\, E_{\mathrm{p}}, \tag{1.4.1}$$

where E_{e} and E_{p} are the beam energies. The nominal beam energies

$$E_{\mathrm{e}} \;=\; 27.5 \text{ GeV}, \qquad E_{\mathrm{p}} = 820 \text{ GeV}$$

result in a center-of-mass energy

$$\sqrt{s} \;\approx\; 300 \text{ GeV}.$$

Defining $\boldsymbol{q}$ to be the photon four-momentum,

$$\boldsymbol{q} \;\equiv\; \boldsymbol{k} - \boldsymbol{k}', \tag{1.4.2}$$

any two of the following three Lorentz-invariant quantities suffice to describe the kinematics:

- the photon virtuality

$$Q^2 \;\equiv\; -q^2 \;=\; Q_{\mathrm{min}}^2 + 4 p_{\mathrm{e}} p_{\mathrm{e}}' \sin^2 \theta_{\mathrm{e}}/2, \tag{1.4.3}$$

with

$$Q_{\mathrm{min}}^2 \;=\; 2(E_{\mathrm{e}} E_{\mathrm{e}}' - p_{\mathrm{e}} p_{\mathrm{e}}' - m_{\mathrm{e}}^2), \tag{1.4.4}$$

where θ_{e} is the electron scattering angle with respect to its initial direction, E_{e}' is the energy of the final-state electron, and the magnitudes of the electron initial- and final-state momenta are designated as p_{e} and p_{e}';
- the Bjorken scaling variable x

$$x \;\equiv\; \frac{Q^2}{2\, \boldsymbol{P} \cdot \boldsymbol{q}}; \tag{1.4.5}$$

- and the Bjorken scaling variable y

$$y \;\equiv\; \frac{\boldsymbol{P} \cdot \boldsymbol{q}}{\boldsymbol{P} \cdot \boldsymbol{k}} \;\approx\; 1 - \frac{E_{\mathrm{e}}'}{E_{\mathrm{e}}} \cos^2 \theta_{\mathrm{e}}/2. \tag{1.4.6}$$

These three variables are simply related in the approximation that the masses are negligible compared to the momenta:

$$Q^2 \;\approx\; s x y. \tag{1.4.7}$$

The minimum kinematically allowed value for Q^2, Q_{min}^2, can be written in the approximation that the electron mass is neglected with respect to its momentum as follows:

$$Q^2_{\min} \approx m^2_e \frac{y^2}{(1-y)}, \tag{1.4.8}$$

and does not exceed 10^{-8} GeV^2 in the kinematic region of the measurements considered in this review. We will discuss data samples for exclusive vector-meson production characterized by three Q^2 ranges:

$$Q^2_{\min} < Q^2 < 4 \text{ GeV}^2 \qquad 0.2 < Q^2 < 0.9 \text{ GeV}^2 \qquad Q^2 > 1 \text{ GeV}^2$$
$$\langle Q^2 \rangle \approx 0.01 - 0.1 \text{ GeV}^2 \quad \langle Q^2 \rangle \approx 0.3 - 0.6 \text{ GeV}^2 \quad \langle Q^2 \rangle \approx 2 - 20 \text{ GeV}^2,$$

corresponding to photoproduction studies, where the final-state electron is not detected, studies at intermediate values of Q^2, where the electron is detected at small scattering angles in a special-purpose calorimeter (the ZEUS BPC), and studies where the electron is detected in the main detector and the upper limit on Q^2 is due simply to luminosity limitations. The values for the average Q^2 cited for the untagged photoproduction studies are rough estimates based on the Q^2 dependence of the photon flux. The corresponding *median* Q^2 value is approximately 10^{-4} GeV^2.

Two further Lorentz-invariant variables which will turn out to be useful in our discussions of the HERA measurements are

$$\nu \equiv \frac{\boldsymbol{P} \cdot \boldsymbol{q}}{m_p} = \frac{Q^2}{2\,x\,m_p}, \tag{1.4.9}$$

which is the value of the photon energy in the proton rest frame, and

$$W^2 \equiv (\boldsymbol{q} + \boldsymbol{P})^2 = -Q^2 + 2\nu m_p + m^2_p \approx sy - Q^2$$
$$\approx Q^2 \frac{1-x}{x}, \tag{1.4.10}$$

the squared center-of-mass energy of the photon–proton system. The kinematic limit for the value of ν is the electron energy in the proton rest frame:

$$\nu_{\max} \approx \frac{s}{2m_p} \approx 48 \text{ TeV}, \tag{1.4.11}$$

and the Bjorken y variable can be written

$$y = \frac{\nu}{\nu_{\max}}. \tag{1.4.12}$$

In the context of the diffractive production of vector mesons, the negative squared momentum transfer at the proton vertex, the Mandelstam variable t, is essential to our discussions of interaction size (see Sect. 2.1). It is defined as follows:

$$t \equiv (\boldsymbol{P} - \boldsymbol{P}')^2 = (\boldsymbol{V} - \boldsymbol{q})^2, \tag{1.4.13}$$

where $\boldsymbol{P}'$ is the four-momentum of the final-state proton and $\boldsymbol{V}$ that of the vector meson. Neglecting the proton mass with respect to its momentum we can write

$$-t \;\approx\; \boldsymbol{p}_{\mathrm{T}}^2(P') = (\boldsymbol{p}_{\mathrm{T}}(V) - \boldsymbol{p}_{\mathrm{T}}(\gamma))^2, \tag{1.4.14}$$

where $p_{\mathrm{T}}^2(P')$ is the transverse momentum of the final-state proton, and $\boldsymbol{p}_{\mathrm{T}}(V)$ and $\boldsymbol{p}_{\mathrm{T}}(\gamma)$ are the transverse momenta of the vector meson and photon. The value of $|t|$ deviates from the squared magnitude of the vector-meson transverse momentum by less than Q^2. The minimum value of $|t|$ kinematically necessary to the elastic process is

$$|t|_{\min} \;=\; m_{\mathrm{p}}^2 \, \frac{(M_{\mathrm{V}}^2 + Q^2)^2}{W^4} \approx \frac{(M_{\mathrm{V}}^2 + Q^2)^2}{4\,\nu^2}, \tag{1.4.15}$$

where M_{V} is the mass of the vector meson, and does not exceed 10^{-4} GeV^2 for the data sets considered in this review.

In concluding our discussion of the kinematics, we introduce the elasticity z, a variable important to the definition of the physical process central to this review. The elasticity is defined as

$$z \;=\; \frac{\boldsymbol{P} \cdot \boldsymbol{V}}{\boldsymbol{P} \cdot \boldsymbol{q}}. \tag{1.4.16}$$

Using the fact that the sum of the squared particle masses is equal to the sum of the Mandelstam variables in the reaction $\gamma^*\mathrm{p} \to V\mathrm{p}$, one can show

$$z \;=\; 1 + \frac{t}{2m_{\mathrm{p}}\nu}. \tag{1.4.17}$$

Thus for values of $|t|$ negligible with respect to $2m_{\mathrm{p}}\nu$, the process we refer to in this review as "elastic" or "exclusive", shown in Fig. 1.3a, restricts the value of the elasticity to unity. Figure 1.3b shows the process accompanied by dissociation of the proton, which results in elasticity $z \lesssim 1$. This process is the primary source of background in the investigations of the elastic process at HERA, except for those in which the final-state proton is detected. In this

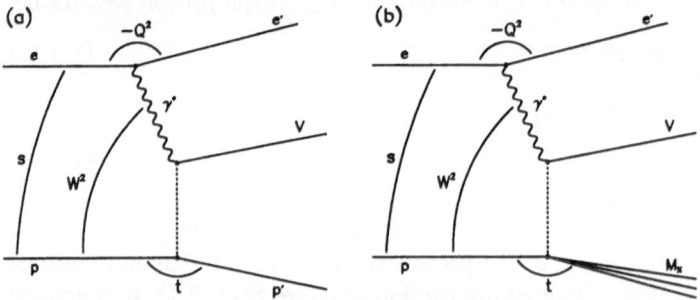

Fig. 1.3. Schematic diagrams for the diffractive electroproduction of single vector mesons: (**a**) elastic production, the principal topic of this review, and (**b**) production with proton dissociation into a hadronic state of invariant mass M_X, one of the primary backgrounds. The dashed line indicates the Pomeron exchange associated with the diffractive process. The definitions of the kinematic variables shown are given in the text

review we will mention only briefly inelastic vector meson production, which is characterized by substantially lower values for the elasticity.

The high photon energy for which data at HERA have been obtained, spanning two orders of magnitude from 400 to 40 000 GeV in the proton rest frame, is precisely the feature of HERA which makes it attractive as a laboratory for photoproduction studies. The high center-of-mass energies in the photon–proton system (40–200 GeV) allow an unprecedented sensitivity to the energy dependence of γ^*p interactions. Equivalently, formerly unmeasured regions of high Q^2 and of low x have now been made accessible, which brings us to our first physics topic: the proton structure functions.

1.4.2 Proton Structure Functions

One of the primary contributions from investigations at HERA to date has been the observation of a strikingly strong x dependence in the proton structure function $F_2(x,Q^2)$ in the hitherto unmeasured kinematic region $x < 10^{-3}$. Figure 1.4 shows the kinematic region covered in measurements performed by the H1 and ZEUS collaborations at HERA in 1993 and 1994, compared to a region covered by ZEUS studies using data from runs in 1995 where the e^+p interaction vertex was shifted by 67 cm in the forward direction (SVTX), and to a region covered by data where the scattered electron was detected in the beam-pipe calorimeter in the ZEUS experiment. Also shown are the kinematic domains in which the proton structure function was measured in fixed-target neutrino, electron, and muon experiments.

For the kinematic domain covered by the HERA measurements, the differential neutral-current cross section is related to the structure functions $F_1(x,Q^2)$, $F_2(x,Q^2)$, and $F_3(x,Q^2)$ as follows:

$$\frac{d^2\sigma^{e^\pm p}}{dx dQ^2} = \frac{2\pi\alpha^2}{xQ^4} \left[2(1-y)\, F_2 + y^2\, 2xF_1 \mp (2y - y^2)\, xF_3 \right] \left[1 + \delta_r \right]$$

$$(1.4.18)$$

$$= \frac{2\pi\alpha^2}{xQ^4} \left[(2 - 2y + y^2)\, F_2 - y^2 F_L \mp (2y - y^2)\, xF_3 \right] \left[1 + \delta_r \right],$$

$$(1.4.19)$$

with

$$F_L(x, Q^2) = F_2(x, Q^2) - 2xF_1(x, Q^2), \qquad (1.4.20)$$

and where $\delta_r(x, Q^2)$ represents a small correction for electroweak radiative effects. The contribution of the parity-violating part from Z^0 exchange, $F_3(x,Q^2)$, does not exceed 1% for $Q^2 < 10^3$ GeV^2, and can be reliably estimated using the expectations of the electroweak Standard Model. The Z^0 contribution to F_2 is similarly small and can be corrected for in the same

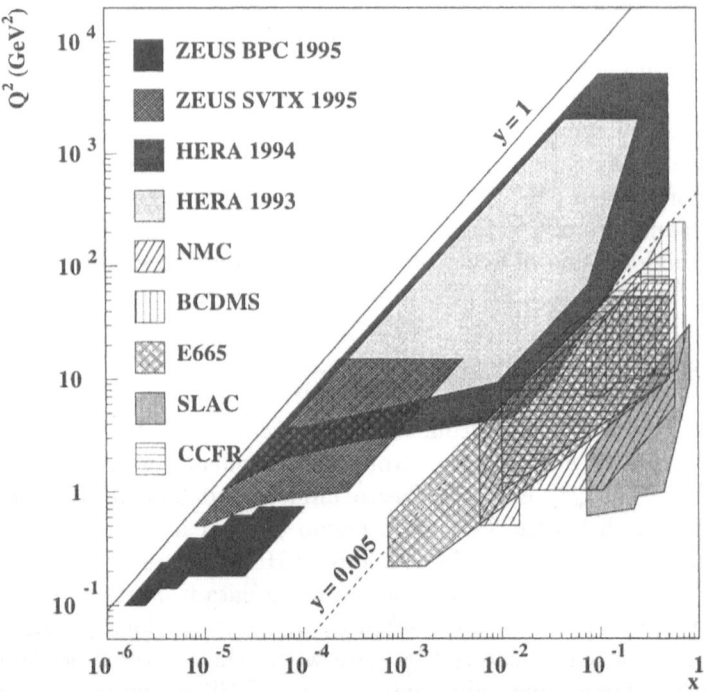

Fig. 1.4. The kinematic region covered by the measurements of the proton structure function $F_2(x,Q^2)$ by the H1 and ZEUS experiments at HERA in 1993 and 1994, compared to a region covered by ZEUS studies using data from runs in 1995 where the e^+p interaction vertex was shifted by 67 cm in the forward direction (SVTX), and to a region covered by data where the scattered electron was detected in a beam-pipe calorimeter (BPC) in the ZEUS experiment. Also shown are the kinematic domains in which the proton structure function was measured in the fixed-target muon experiments NMC and BCDMS at CERN, and E665 at FNAL. The region covered by the fixed-target neutrino experiment CCFR at FNAL is shown, as well as that covered in the early SLAC fixed-target studies with electron beams

manner. Thus the differential cross section is primarily determined by the value of the electromagnetic part of F_2, which we call F_2^{em}:

$$\frac{\mathrm{d}^2\sigma^{e^\pm p}}{\mathrm{d}x\mathrm{d}Q^2} = \frac{2\pi\alpha^2}{xQ^4}\left[2(1-y) + \frac{y^2}{1+R}\right] F_2^{\mathrm{em}} \left[1 + \delta_Z\right] \left[1 \mp \delta_3\right] \left[1 + \delta_r\right],$$

$$(1.4.21)$$

with R defined as

$$R(x, Q^2) = \frac{F_L(x, Q^2)}{2xF_1(x, Q^2)} = \frac{F_2(x, Q^2)}{2xF_1(x, Q^2)} - 1, \qquad (1.4.22)$$

and the small positive values of $\delta_3(x, Q^2)$ and $\delta_Z(x, Q^2)$ are the aforementioned corrections. The value of R contributes corrections of order 10% at values of $y \lesssim 1$ and can be calculated in the context of perturbative quantum chromodynamics (pQCD) [25].

The H1 and ZEUS collaborations have published a series of measurements of $F_2(x,Q^2)$ in the range $3.5 \times 10^{-5} < x < 3.2 \times 10^{-1}$ and $1.5 < Q^2 < 5.0 \times 10^3$ GeV^2 [26, 27, 28, 29]. Figure 1.5 shows an example of these measurements featuring the strong x dependence and compares them to the measurements with muon beams at higher x performed by the NMC collaboration [30]. Also shown is the result of a fit [31] to the data based on a next-to-leading-order (NLO) approximation to the Dokshitzer–Gribov–Lipatov–Altarelli–Parisi (DGLAP) [32] evolution equations which describe the dependence on Q^2. Such a dependence of F_2 on x has long been the subject of theoretical interest. In 1974 De Rújula et al. [33] predicted a dependence growing with x^{-1} more weakly than a power law but more rapidly than any power of its logarithm. López et al. [34] have discussed the possibility of a power-law dependence. In Sect. 3.1 we will see how this x dependence is related to the energy dependence in the total photon–proton cross section.

While the photon does not couple to the gluons at the Born level, its interactions with the quarks and antiquarks at low x result in sensitivity to the gluon momentum distribution. Both the H1 and ZEUS collaborations have derived gluon densities from their measurements of the proton structure function [35, 36]. The structure function F_2 can be decomposed into singlet and nonsinglet parts. The latter is insensitive to the gluon content of the proton and its Q^2 dependence differs from that of the singlet part. The singlet part depends on the gluon content and its evolution with Q^2 allows the extraction of the gluon momentum density $xG(x, Q^2)$. Figure 1.6 shows the results for this density obtained from a global NLO QCD fit by the H1 collaboration to their 1994 data. A steep decrease with x is evident within the margin of uncertainty, as the density function decreases by a factor of approximately four over the measured region of x for $Q^2 = 20$ GeV^2. We will see that the gluon density function plays a crucial role in the interpretations of elastic vector-meson production at HERA based on perturbative QCD.

We close this brief review of HERA studies of the structure of the proton by citing the first presentation of a result for the longitudinal structure function $F_L(x,Q^2)$ [37]. The H1 collaboration has used the F_2 measurements at low y, where the contribution from F_L is small, to extract parton densities, and then extrapolated these results to estimate F_2 at high y and to isolate the contribution of F_L, obtaining the result $F_L = 0.52 \pm 0.03$ (stat)$^{+0.25}_{-0.22}$ (sys) at $x = 2.43 \times 10^{-4}$ and $Q^2 = 15.4$ GeV^2. The extrapolation algorithm presumes the validity of the DGLAP evolution procedure. The result is consistent with a QCD calculation which employs the gluon density function determined from the measurement of F_2. The experimental challenge of accurately measuring the energy of the scattered electron in the range $6 < E'_e < 11$ GeV at high y is evinced by the systematic uncertainties in the result.

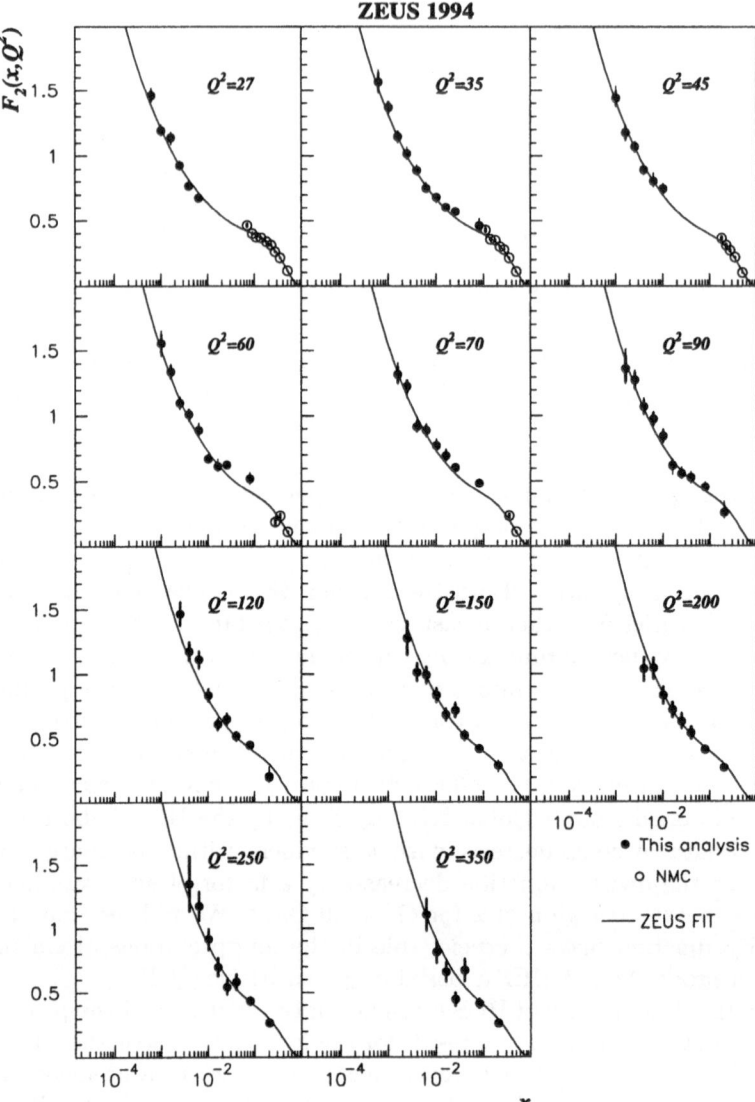

Fig. 1.5. An example of measurements by the ZEUS collaboration of the structure function $F_2(x,Q^2)$, shown as the solid circles, and the results of a QCD NLO fit to the data [29]. The Q^2 values are given in units of GeV^2. The error bars represent the sum in quadrature of the systematic and statistical uncertainties. The open circles show measurements obtained by the NMC collaboration in fixed-target studies with a muon beam [30]

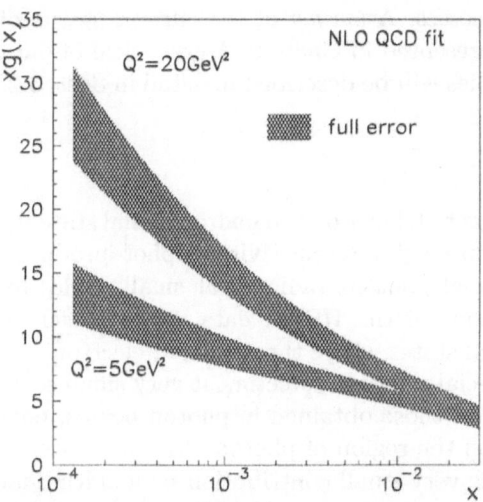

Fig. 1.6. Results for the gluon momentum density $xG(x, Q^2)$ determined by the H1 collaboration using NLO QCD fits to their 1994 measurements of the proton structure function F_2 [35]. The shaded bands indicate the region of uncertainty in the result for the gluon density, obtained by quadratic addition of systematic uncertainties with those due to the uncertainties in the fit parameters determined via the minimization procedure

1.4.3 Hadronic Final States in Deep Inelastic Scattering

The electron–proton center-of-mass system at HERA moves in the laboratory reference frame in the proton flight direction with a value for the Lorentz factor γ equal to 2.8. The unique laboratory reference frame provided by the advent of these asymmetric storage rings in the study of deep inelastic scattering results in good acceptance for the scattered proton constituent and thus opens a vast field of study of hadronic final states. The high center-of-mass energy, furthermore, extends the accessible rapidity range and results in more collimated, and hence more easily identified, jets in the final state. A wide range of investigations have been performed, including dijet production, the measurement of α_s from multijet rates, energy flows, inclusive charged particle distributions, strangeness production, charm production, and much more [38, 39]. The early investigations of event topologies in deep inelastic scattering revealed a class of events with no energy flow in the forward region, amounting to approximately 10% of the total observed rate. Thus a clean signal for diffractive physics at high photon virtuality was obtained, opening the way to a variety of studies ranging from inclusive transverse momentum distributions and energy flows to studies of Pomeron structure [39, 40, 41].

The investigations of exclusive final states in deep inelastic scattering have now become of topical theoretical interest, as diffractive vector-meson production at high Q^2 has been recognized to yield information on the relative contributions of hard and soft processes. A survey of the current status of these theoretical considerations is presented in Chap. 2. A great deal of data has become available and these studies will be described in detail in Sect. 3.2.

1.4.4 Photoproduction

The ability of the HERA experiments to trigger on the hadronic final state independently of the scattered electron results in sensitivity to photoproduced final states provided by the quasi-real photons radiated at small angles by the electron beam. In our discussions of the HERA data we will refer as "photoproduction" to studies of final states where the scattered electron was either undetected or detected in special-purpose detectors at very small scattering angle, comparing the results to those obtained in photon beams, basing the validity of the comparison on the region of photon virtuality covered ($\langle Q^2 \rangle \approx 0.01 - 0.1\,\mathrm{GeV}^2$) and on the very small contribution by longitudinal photons to the photoabsorption cross section compared to that by transverse photons at such low values of Q^2. The HERA data extend the range of accessible photon energies by two orders of magnitude compared to that achieved in fixed-target experiments.

The large photoproduction cross section at HERA, three orders of magnitude higher than that for deep inelastic processes where the scattered electron is observed in the central detectors, has resulted in the availability of high-statistics investigations of photon interactions and hadronic final states characterized by transverse energies far exceeding the QCD scale parameter Λ_{QCD}. An overview of photoproduction physics topics addressed by the first three years of HERA operation can be found in [42]. A general review of theoretical issues in photoproduction is given in [43]. Reference [44] reviews the status of the HERA investigations of the partonic structure of the photon. The H1 and ZEUS collaborations have published numerous studies of final-state jet topologies and inclusive charged particle rapidity distributions and energy flows, quantified the occurence of rapidity gaps between jets, and distinguished contributions from direct and resolved photon interactions [45, 46]. Diffractive processes have been identified via the presence of a rapidity gap with respect to the proton direction, resulting in a number of studies of diffractive hadronic final states, including the observation of hard scattering [46, 47]. Cross sections for the photoproduction of charmed mesons have been determined [48]. The measurements of total photon–proton cross sections and vector-meson photoproduction are essential to the topic of this review and will be discussed in detail in Sect. 1.5.2.

1.5 Vector-Meson Production in Photon–Proton Interactions

1.5.1 Experimental Techniques Prior to HERA Operation

A comprehensive description of vector-meson photoproduction and electro-production experiments prior to 1978 is available in the review by Bauer et al. [49]. A wide variety of photoproduction studies on hydrogen targets were performed at the Cambridge Electron Accelerator, the Wilson Synchrotron Laboratory at Cornell, SLAC, and DESY. Photon beams of energies between 2 and 18 GeV were obtained via electron-beam bremsstrahlung and via backscattered laser beams. Hybrid bubble chamber studies at SLAC [50] and streamer chamber studies at DESY [51] permitted complete reconstruction of the final state in many cases, while wire chamber spectrometers of relatively limited acceptance detected either the proton alone, reconstructing the vector meson via missing-mass techniques, or the vector-meson decay products, or both, in two-arm setups. Studies of ρ^0 mesons dominate the statistics, and asymmetries in the dipion mass spectrum generated much theoretical interest, as did the helicity analyses of the decay angle distributions. Evidence for the interference of the ρ^0 and ω production amplitudes was observed [52] in the dipion invariant mass spectra, which also exhibited further asymmetries in the ρ^0 resonance region attributed to interference with nonresonant backgrounds. Photoproduction on protons of the ω, ϕ, and ρ' was studied [53]. Electro- and muoproduction of the ρ^0, ω, and ϕ mesons up to Q^2 values of about 2 GeV^2 was investigated at DESY, SLAC, and Cornell [51, 54]. High-energy muoproduction of the ρ^0 meson was studied by the CHIO collaboration at FNAL [55] and by the EMC [56, 57] and NMC [58, 59] collaborations at CERN. Recently results for ρ^0 muoproduction from the FNAL experiment E665 have also become available [60], reaching a virtual photon energy of 420 GeV. Diffractive ρ^0 production has also been investigated in bubble chamber studies of neutrino and antineutrino interactions with protons [61]. Soon after their discovery, the J/ψ and $\psi(2S)$ mesons were observed in photoproduction at Cornell [62], FNAL [63], and SLAC [64], the FNAL measurements extending the photon energy range to 100 GeV. Photoproduction of charm states prior to 1985 is covered in the review by Holmes et al. [65]. The production of J/ψ and $\psi(2S)$ mesons was studied using fixed-target spectrometers in tagged and wideband photon beams for photon energies up to 225 GeV at FNAL [66]. Comparable energies were reached in muoproduction by the Berkeley/Princeton/FNAL collaboration [67] at FNAL and by the EMC [68] and NMC [69] collaborations at CERN. The NA14 collaboration at CERN also measured J/ψ photoproduction in a CERN photon beam at such energies [70]. In more recent years the fixed-target experiment E687 in the photon beam at FNAL has published results for J/ψ photoproduction [71].

1.5.2 The Contribution from Investigations at HERA

This section briefly summarizes the contributions to this field of research presented by the two collaborations H1 and ZEUS prior to the end of 1996. The range of forthcoming studies at HERA to be expected in the coming years will be covered in Chap. 4. The section entitled "Suggestions for Future Work" in the review of Bauer et al. [49] presents an instructive comparison to these issues.

The measurements at HERA we discuss in this review concern a range of photon energies in the proton rest frame extending from 400 GeV to 40 TeV. Of immediate consequence are the small lower kinematic bounds on photon virtuality, Q^2_{min}, and momentum transfer at the hadronic vertex (see Sect. 1.4.1), rendering their influence on the determination of kinematical variables negligible for most studies. The variety of diffractive processes which contribute coherently to the interaction with the proton increases with decreasing $|t|_{min}$ [72]. Thus the kinematic region covered at HERA allows an unprecedented sensitivity to the partonic structure of the photon in its diffractive interactions. In contrast to the fixed-target studies, the HERA experiments study both photoproduction and electroproduction processes simultaneously, distinguishing them via selection algorithms requiring either a small energy deposit in the main calorimeters and at least one charged particle track or a large electromagnetic energy deposit signalling the detection in the central detector of an electron scattered with large momentum transfer.

The present HERA results also extend the range of accessible photon virtualities by more than two orders of magnitude, reaching Q^2 values of 5×10^3 GeV^2. Thus the energy dependence of the total photon–proton cross section for highly virtual photons has been studied at high energy for the first time, and comparisons to the studies at low energy provide high sensitivity even to the very weak energy dependences expected in diffractive processes. Particular attention has been given to the intermediate region of photon virtuality $0.1 < Q^2 < 1.0$ GeV^2, where the transition from vector dominance to the region of applicability of perturbative quantum chromodynamical calculations can be studied.

Investigations of photoproduction of ρ^0, ω, ϕ, J/ψ, and $\psi(2S)$ mesons have been published. In addition results on the diffractive production of the ρ^0, ρ', ϕ, and J/ψ mesons at high Q^2 have been presented. The energy dependence of the forward ($t = 0$) cross sections has been studied via comparison with measurements at low energy. Production ratios have been quantified in this high-energy region where, for example, the rates of ω and ϕ production are expected to become comparable since the contributions to ω production other than Pomeron exchange become negligible. The dependence on the momentum transfer at the proton vertex t has been studied up to values of $|t|$ of about 4 GeV^2. Helicity analyses of the vector-meson decay products have been presented, leading to estimations for the ratio of longitudinal to transverse photon cross sections. This variety of measurements now available has prompted a great deal of new theoretical work, which we summarize in the following section.

2. Contemporary Theoretical Approaches

A broad array of theoretical considerations has been brought to bear on the problem of exclusive vector-meson production at HERA over the past few years. The high level of interest has been stimulated in part by the wide kinematic range covered by the data, with photon virtualities ranging from 10^{-10} to 30 GeV^2 and photon–proton center-of-mass energies from 40 to 200 GeV. The range of vector-meson masses for which measurements have become available, extending from that of the ρ^0 to that of the J/ψ, provides a further parameter to be used in testing theoretical ideas. The photoproduction data show features characterizing diffractive processes and measurements at high Q^2 have yielded strong evidence for the applicability of approaches based on perturbative QCD calculations. Thus the HERA measurements are demanding the reconciliation of theoretical ideas which have hitherto been tested only in distinct kinematic domains. The purpose of this chapter is to summarize the current status of these theoretical approaches. Following a brief discussion of model-independent issues, we summarize a variety of topical nonperturbative calculations and end with a discussion of the status of calculations based on perturbative quantum chromodynamics.

2.1 General Considerations

The early measurements at HERA of total photon–proton cross sections and of ρ^0 meson production exhibited characteristics typical of soft diffractive processes. The magnitudes of the cross sections were comparable to those measured at an order of magnitude lower energy, and low momentum transfers t were strongly favored. In the brief outline of the general characteristics of such processes we follow the discussion of [73]. A beautifully detailed and comprehensive discussion of the Regge theory which successfully describes these phenomena can be found in [74].

The strong forward peaking of the diffractive processes mentioned above refers to an exponential t dependence in the region of small $|t|$:

$$\frac{d\sigma}{d|t|} = Ae^{-b|t|} \tag{2.1.1}$$

In terms of a simple optical model interpretation, the exponential slope b is related to the radii of two Gaussian-shaped objects participating in the elastic scatter:

$$b \ \propto \ R_1^2 + R_2^2. \tag{2.1.2}$$

This review discusses the HERA results on elastic scattering from the proton, for which the relevant squared radius has been determined to be near $4 \ \mathrm{GeV}^{-2}$ in its hadronic interactions, so the slopes discussed in this review yield information on the size of the object scattering on the proton, usually interpreted as a $q\bar{q}$ wave packet.

The optical theorem relating the imaginary part of the forward scattering amplitude to the total cross section is useful here, since for purely diffractive elastic processes the real part is zero and measurements show the imaginary part exceeding the real part by at least a factor of three even at low energies. At HERA energies the real part is less than a few percent and we can relate the above differential cross section to the total cross section as follows:

$$\left. \frac{\mathrm{d}\sigma}{\mathrm{d}|t|} \right|_{t=0} = A = \frac{\sigma_{\mathrm{tot}}^2}{16\pi}. \tag{2.1.3}$$

If we now integrate the differential cross section to get the total elastic cross section and express the cross sections in millibarns and the exponential slope b in units of GeV^{-2} we obtain

$$\sigma_{\mathrm{el}} \ = \ \frac{0.051 \, \sigma_{\mathrm{tot}}^2}{b}. \tag{2.1.4}$$

In this manner it becomes obvious that the energy dependences of the elastic and total cross sections are closely related by that of the exponential slope b. Such an energy dependence, termed "shrinkage" since the width of the forward diffractive peak in the t distribution decreases with energy, has been accurately measured in low-energy hadronic interactions and found to vary starkly with the process considered. Of uncertain dynamical origin, this effect remains topical today in the investigations of the HERA data.

The energy dependence of the forward diffractive cross section derived from Regge theory follows a power law, where the power is determined by the exchanged trajectory:

$$\frac{\mathrm{d}\sigma}{\mathrm{d}|t|} \propto \mathrm{e}^{-b_0|t|} \left(\frac{s}{s_0} \right)^{2(\alpha(t)-1)}. \tag{2.1.5}$$

The energy scale factor s_0 is determined empirically to be comparable to the hadronic mass scale of about $1 \ \mathrm{GeV}^2$. The t dependence of the linear trajectory,

$$\alpha(t) \ = \ \alpha_0 + \alpha' t, \tag{2.1.6}$$

leads directly to the aforementioned shrinkage effects,

$$b \ = \ b_0 + 2\alpha' \ln \frac{s}{s_0}, \tag{2.1.7}$$

and an energy dependence of the total cross section determined by the $t=0$ intercept of the trajectory (see (2.1.4))

$$\sigma_{\text{tot}} \propto \left(\frac{s}{s_0}\right)^{(\alpha_0-1)}. \tag{2.1.8}$$

The weak energy dependence of the total scattering cross sections measured at low energy motivated the introduction of the Pomeron trajectory with intercept near unity, and these considerations were again brought to the fore when the early HERA results for the total photoproduction cross sections gave results quantitatively comparable to the measurements at low energy.

We conclude this general discussion of diffractive processes with a remark clarifying a convention of terminology. Since the exclusive reactions $\gamma^*p \rightarrow Vp$ we discuss in the review involve more final-state rest mass than in the initial state, these reactions are not, strictly speaking, "elastic" interactions. However, the recent literature has adoped the term "elastic" for these processes. The reason is based on kinematics: the minimum momentum transfer required by the generation of the final-state rest mass is negligibly small compared to typical values of the momentum transfer measured at the high center-of-mass energies of the HERA regime. The reactions we consider are called "diffractive inelastic" in [73], for example, while today's theoretical and experimental articles reserve the term "inelastic" for processes in which the photon or proton dissociates into several final-state particles. Thus an essential difference between diffractive elastic and diffractive inelastic processes discussed in [73], namely the conservation of intrinsic spin and parity in the elastic interactions, becomes moot in the context of the present review, since in the helicity analyses of our "elastic" reactions we investigate the possibility of spin and parity changes in order to gain information on the dynamical origin of the interaction.

We continue our discussion of general theoretical considerations by turning to the issue of vector-meson production ratios. The quark-flavor structure of the vector mesons, together with the assumption of a flavor-independent production mechanism governed by the electromagnetic quark–photon coupling in the quark–parton model, results in production rates with the relationship

$$\rho^0 : \omega : \phi : J/\psi = 9 : 1 : 2 : 8. \tag{2.1.9}$$

This flavor symmetry can be expected to be broken by effects related to the vector meson masses. Indeed, a suppression factor of about a factor of three was observed in elastic ϕ photoproduction at low energies [49] and for J/ψ photoproduction [65] this suppression was shown to be more than two orders of magnitude. On a model-independent basis it is reasonable to expect the symmetry to be restored for $Q^2 \gg M_V^2$. Further discussion of expectations for the production ratios at high Q^2 can be found in [72].

General arguments favoring the dominance of vector-meson photoproduction by longitudinal photons over that by transverse photons at high Q^2 have been made based on the point-like nature of the longitudinal photon. The resulting factorization theorem and dimensional counting considerations [75] result in a prediction for the Q^2 dependence of the γ^*p $\rightarrow$ Vp cross section to vary as Q^{-6}. However, there are no model-independent arguments for the Q^2 dependence of the ratio of the two cross sections (aside from the fact that the ratio must vanish at $Q^2 = 0$ owing to gauge invariance), complicating the comparison with the HERA data, which comprise approximately equal contributions from transverse and longitudinal photons.

If helicity is conserved in the photon/vector meson transition, then the ratio of the longitudinal to transverse cross sections can be determined by measuring the vector meson polarization via its decay angle distributions. This transition poses a fundamental quandary, however, and questions concerning the structure of the vector-meson spin-density matrix have played a prominent role thoughout the history of investigations into exclusive vector-meson photo- and leptoproduction. The quandary arises from the rest mass of the vector meson, which results in an ambiguity in the choice of Lorentz frame appropriate to a description of the underlying dynamics. Early studies at low energy and low Q^2 employed comprehensive helicity analyses to show that a frame could indeed be defined in which the spin-density matrix elements were independent of the kinematics and in which helicity was conserved. It is important to bear in mind while reading this review, however, that no such analysis has yet been possible with the HERA data, and conclusions concerning the ratio of longitudinal to transverse cross sections are based on an assumption motivated by measurements in a kinematic region not covered by the HERA data. A detailed description of the spin-density matrix, which describes the helicity structure of the interaction, will be presented in Sect. 3.2.5.

Summarizing these general theoretical considerations, the interpretation of the HERA photoproduction results for the lower-mass vector mesons as being of soft diffractive origin has been quite successful. However, the measurements of J/ψ photoproduction at high energy as well as the exclusive production of vector mesons at photon virtualities greater than 7 GeV^2 show starkly different energy dependences. Model-independent theoretical considerations point to the need for explanations based on perturbative quantum chromodynamics. The mass of the J/ψ introduces a hard scale which casts doubt on the applicability of calculations based on single soft Pomeron exchange. And on general grounds one expects the longitudinal component of the virtual photon to dominate the transverse component at high Q^2, so it is plausible to expect considerations based on the optical model to lose their validity. Indeed the experimental results exhibit energy dependences incompatible with those derived from calculations based on soft diffraction. These considerations have made the exclusive production of vector mesons in high-energy γ^*p interactions the subject of lively theoretical interest from experts in both nonperturbative and perturbative approaches to the description of hadrons and their interactions. In the following two sections we briefly summarize the recent theoretical work in this field.

2.2 Nonperturbative Approaches

Donnachie and Landshoff [76] have employed the highly successful parametrizations of the hadron–hadron total cross sections [77], based on the exchange of an object we will refer to as the phenomenological Pomeron, along with an assumption of vector-meson dominance (VMD) [49, 78, 79] and the additive quark model to describe vector-meson photoproduction. They used the e^+e^- decay width of the ρ^0 to estimate the $\gamma^*\rho^0$ coupling. The Regge-theory-inspired fits to the hadronic cross sections yield values for the Pomeron trajectory of

$$\alpha_{\mathbb{P}}(t=0) \quad \approx \quad 1.08 \qquad \alpha'_{\mathbb{P}} \approx 0.25 \ \mathrm{GeV}^{-2}, \qquad (2.2.10)$$

resulting in the expectation of a power-law increase of the elastic cross section with energy, the value of the power being less than $4(\alpha_{\mathbb{P}}(t=0)-1) \approx 0.32$, depending on the t range covered by the data (see (2.1.4) and (2.1.8)), and in the expectation of shrinkage. They furthermore suggested an extension of these considerations to high Q^2, modeling the Pomeron as a pair of nonperturbative gluons, predicting a Q^{-6} dependence at high Q^2.

The model of **Haakman et al.** [80] derived an effective intercept based on the hypothesis of multiple Pomeron exchanges at high energies. This intercept leads to a stronger energy dependence and successfully matches the HERA results for the energy dependence in the total γ^*p cross section as well as in exclusive ρ^0 and J/ψ photoproduction and in ρ^0 production at high Q^2.

Jenkovszky et al. [81] have addressed the energy dependence of J/ψ photoproduction. They succeed in explaining the steep energy dependence within the framework of single Pomeron exchange (indeed, the "pure" Pomeron of unity intercept), attributing the steep energy dependence to preasymptotic effects arising from the high J/ψ mass. This high mass alters the relationship between the elastic and inelastic contributions to the total cross section, moving the contribution from the elastic cross section to higher energy compared to that from the elastic production of the low-mass vector mesons.

Pichowsky and Lee [82] have discussed electroproduction of ρ^0 mesons in terms of a Pomeron-exchange model where the ρ^0–Pomeron coupling is represented by a nonperturbative quark loop, predicting the Q^2 dependence of the elastic cross section in the region of transition to the perturbative regime $0.1 < Q^2 < 10 \ \mathrm{GeV}^2$, and successfully matching results from the fixed-target muon experiments. The energy dependence of the elastic cross section in this model is weak, similar to that for the phenomenological Pomeron exchange.

The nonperturbative calculations of **Dosch et al.** [83] used a stochastical model of the QCD vacuum yielding linear confinement to derive high-energy vector-meson leptoproduction cross sections for each of the vector mesons, based on hadron sizes. They found t dependences which are less steep than exponential and Q^2 dependences which do not reach the asymptotic behavior of Q^{-6} even for Q^2 values as high as 10 GeV^2. Of particular interest are their predictions for the Q^2 dependence of the ratio of longitudinal to transverse cross sections, which show a much weaker dependence than the proportionality to Q^2/M_V^2 expected on the basis of vector-meson dominance, and which match the currently available HERA measurements of this ratio. However, this model has no prediction for the energy dependence of the elastic cross sections.

Niesler et al. [84] have published a calculation of ρ^0 leptoproduction which derives the $M_{\pi^+\pi^-}$ spectrum from the two-pion contribution to the photon spectral function, which is given by the pion form factor. The coupling of the ρ^0 meson to the nonresonant continuum was understood in terms of an effective field theory approximating QCD for composite hadrons. The calculation yielded good agreement to a measurement of the mass spectrum at low energy where a cusp arising from ρ^0–ω mixing is evident. Consistency with the ZEUS photoproduction result [85] was also achieved, though the statistical accuracy of this early measurement at high energy is much weaker, preventing a clear observation of such an interference cusp, if it is present.

2.3 Calculations Based on Perturbative Quantum Chromodynamics

Shortly before data-taking began at HERA in 1992, **Ryskin** [86] suggested that diffractive J/ψ electroproduction could be calculated in perturbative QCD (pQCD) even at low Q^2, since the J/ψ mass furnished a scale large compared to the confinement scale Λ_{QCD}, thus permitting the perturbative expansion. His calculation, based on the exchange of a momentum-symmetric pair of perturbative gluons, indicated that the cross section is sensitive to the square of the gluon density evaluated at effective values of Q^2 and x:

$$\bar{Q}^2 = \frac{Q^2 + M_{J/\psi}^2}{4} \quad \text{and} \quad \bar{x} = \frac{4\bar{Q}^2}{W^2}. \tag{2.3.11}$$

Interest in this approach was further stimulated when the early experimental results showed both that the proton structure function rises sharply as x decreases and that the energy dependence of the diffractive J/ψ photoproduction cross section is much steeper than expected from a soft diffractive process. The Ryskin calculation has since been implemented in a Monte Carlo simulation package [87]. Some theoretical uncertainties, including the need for the unmeasured two-gluon form factor, mitigate the quantitative use of this process as a measure of parton densities, but hope remains strong that this quadratic dependence will allow high sensitivity to the gluon density. A first attempt to discriminate among parton density parametrizations has been made by **Ryskin et al.** [88], where the authors extended the calculation beyond the leading-log approximation and indeed concluded that the energy dependence of the diffractive J/ψ photoproduction cross section provides a more sensitive probe of the gluon density than the proton structure function measurements.

Brodsky et al. [89] have generalized this type of calculation to the diffractive production of low-mass vector-meson states (indeed any hadronic states with the quantum numbers of the photon), making several distinctive predictions for measurements in the kinematic range of low x, $s \gg Q^2 + M_V^2$, and $Q^2 \gg \Lambda_{QCD}^2$ addressed by measurements at HERA. These included the predominance of the amplitude for longitudinally polarized photons producing longitudinally polarized vector mesons, t slopes independent of the produced vector-meson wave function and flavor content, a slow increase of these t slopes with energy, ratios of the vector-meson production cross sections closely related to standard SU(3) wave function predictions, the Q^{-6} dependence, and the dependence on the square of the gluon density function in the proton. The calculation employed the momentum distribution of constituents in the virtual photon, the two-gluon form factor of the proton depending on the gluon density function $G(x, \bar{Q}^2)$ (which alone determines the t dependence at high Q^2), and the nonperturbative wave function of the final-state vector meson, for which they used the minimal Fock-state wave functions of Brodsky and Lepage [90]. Their results reduce to those of Ryskin when a simple nonrelativistic symmetric pair of quarks is assumed for the vector meson. They showed in a double-leading-log approximation that the longitudinal part of the forward cross section can be written

$$\left.\frac{d\sigma_{\gamma^* p \to V p}}{d|t|}\right|_{t=0} = A_V \frac{\alpha_s^2(\bar{Q}^2)}{\alpha \bar{Q}^6} \left|\left[1 + \frac{i\pi}{2} \frac{d}{d(\ln x)}\right] xG(x, \bar{Q}^2)\right|^2, \quad (2.3.12)$$

where α_s is the strong coupling constant and α the fine structure constant. The additional term proportional to the logarithmic derivative of the gluon momentum density is a perturbation arising from the real part of the amplitude calculated via dispersion relations, and results in an effective QCD Pomeron intercept exceeding unity at small x and high Q^2. The normalization factor A_V is a constant derived from the vector-meson wave function which leads to an uncertainty of approximately a factor of three in the normalization of the predicted cross section for the ρ^0, for example. For the interactions of the longitudinal photons the ratio of the production of longitudinally polarized vector mesons to that of transversely polarized vector mesons was predicted to rise even faster than Q^2/M_V^2. Particularly interesting predictions were also made for diffractive leptoproduction of vector mesons off nuclei, including decisive tests of color transparency, but these will not be tested at HERA until nuclear beams are introduced.

Frankfurt et al. [91] have applied the above ideas in quantitative calculations of the magnitudes, energy dependences, and cross section ratios for vector-meson leptoproduction and compared them to the NMC and HERA

measurements. They discussed pQCD predictions for the vector-meson production ratios at high Q^2, as well as for the production of excited states of
vector mesons and for the diffractive production off nuclei. The kinematic
limits of validity for these perturbative calculations were explicitly quantified. They concluded that the study of exclusive diffractive production of
vector mesons presents an effective method for determining the vector-meson
wave functions and may therefore provide useful tests of QCD lattice gauge
calculations at high energy.

Hoodbhoy [92] has addressed the magnitude of two contributions to
diffractive J/ψ electroproduction neglected in the early calculations: that of
the Fermi motion of the c quarks in the J/ψ and that of asymmetric gluon
pairs in the proton, finding them to be small.

Collins et al. [75] have also succeeded in generalizing the calculations
beyond the leading-log approximation. The calculation was restricted to longitudinally polarized photons, though the contributions by transversely polarized photons were extensively discussed. One interesting conclusion is that
the ratio of longitudinally to transversely polarized vector mesons is sensitive
to the polarized quark densities in the proton, so that information on these
densities can be obtained from diffractive vector-meson production without
requiring polarized proton beams. This remark holds only at high x, however,
since the production of transversely polarized vector mesons is predicted to
vanish at small x. The predicted ratio of longitudinal to transverse vector-
meson polarization is much larger than the value measured at HERA, leading
the authors to surmise that the HERA results have a substantial contribution
from transversely polarized photons.

Martin et al. [93] have addressed this issue of the Q^2 dependence of the
vector meson polarization, showing that not only the above perturbative calculations but also nonperturbative calculations of vector-meson production
by transverse photons predict contributions to the production of transversely
polarized ρ^0 mesons which decrease more rapidly with Q^2 than the early
experimental results from HERA indicate. They proposed a model of open
production of light $q\bar{q}$ pairs and used an assumption of parton–hadron duality
to calculate the ratio of the production cross section for longitudinally polarized ρ^0 mesons to that for transverse polarization. Their results yielded a Q^2
dependence for this ratio which is weaker than linear and is consistent with
the present measurements within the rather large experimental uncertainties.

Nemchik et al. [94, 95] have shown that diffractive leptoproduction of
vector mesons provides a sensitive probe of Pomeron dynamics at high energies and that a measurement of the Balitsky–Fadin–Kuraev–Lipatov (BFKL)
Pomeron [96] intercept will be possible when ρ^0 and J/ψ production cross
sections are available for Q^2 values greater than 100 GeV^2. The generalized
BFKL dynamics were used to make specific predictions for the interaction of
the target nucleon with the color dipole representing the virtual-photon state.
The dipole size was determined by the sum $M_V^2 + Q^2$, allowing a scanning

of the dynamics by observing energy dependences while varying M_V and Q^2. Particularly striking effects were predicted for the energy dependence of the radially excited vector-meson states.

Amundson et al. [97] have used the nonrelativistic QCD factorization formalism of Bodwin et al. [98] to calculate J/ψ forward photoproduction. This calculation results in a linear dependence on the gluon density function. Good agreement with the fixed-target and HERA measurements of the energy dependence of the J/ψ photoproduction cross sections was found at the level of present experimental uncertainties.

Ivanov and Ginzburg [99] have discussed the hard diffractive photoproduction of light mesons and photons for both real and virtual initial-state photons in the approximation that the transverse momentum of the final-state meson or photon exceeds the QCD confinement scale. They quantify the region of pQCD validity for the production of ρ^0 and ϕ mesons and for the hard Compton effect. They emphasize that a signature for this region of validity is the exclusively longitudinal polarization of a final-state vector meson, independent of the initial-state photon polarization, which is a consequence of the chiral nature of perturbative couplings. This review also contains a useful survey of contemporary theoretical work in this field.

Bartels et al. [100] and **Forshaw and Ryskin** [101] have pointed out that the hard scale necessary for the applicability of perturbative calculations can be given by the momentum transfer at the proton vertex t as well as by the photon virtuality and the vector-meson mass. They show that the BFKL equation can be used to predict the t dependence in diffractive vector-meson production in the approximation $|t| \gg \Lambda^2_{\text{QCD}}$. At present the HERA measurements at high t are statistically too weak to permit a quantitative comparison, but sufficient information should become available in coming years to test the calculations.

3. Experimental Results
from Investigations at HERA

Our discussion of the present experimental status of investigations of vector-meson electroproduction at HERA begins with a brief description of the total inclusive cross sections for the interaction of real and virtual photons with the proton. This description will contrast the various physical processes contributing to these cross sections, resulting in a model-independent perception of their dependence on energy and photon virtuality in the kinematic domains covered by the available measurements. Our goal is to maintain enough generality to permit a unified understanding of these interactions, bridging the gaps between the various kinematic regions presently described independently in their corresponding kinematic approximations. We are motivated to such an approach by the broad kinematic range covered by HERA measurements, which beautifully complements earlier investigations of photoproduction processes and of deep inelastic lepton scattering.

3.1 Total Photon–Proton Cross Sections

The use of a high-energy electron beam for investigations of proton structure at HERA has as a direct consequence the dominating contribution of real-photon–proton interactions, referred to as photoproduction processes. The flux of radiated photons in the energy range used for the HERA measurements from such a beam is a factor of 2–3 higher than that from a muon beam of similar energy, for example, and the total cross section for photon–proton interactions is a mere factor of α^{-1} less than the hadron–hadron total cross section, where α is the fine structure constant. This ≈ 100 μb cross section for photoproduction is three orders of magnitude greater than the part of the deep inelastic cross section for which the central H1 and ZEUS detectors accept the scattered electron ($Q^2 \gtrsim 1$ GeV^2). While for photoproduction processes the scattered electron exits the central detector via the rear beam pipe, any part of the rest of the final state which mimics an electron causes a serious background to studies of deep inelastic scattering and limits the kinematic region accessible for its accurate measurement. This high rate for photoproduction processes led directly to the first publications of physics results from the first HERA running in June of 1992 [102, 103]. Each experiment repeated the measurement with the higher statistics available in later data runs, further attempting to quantify the various elastic and dissociative subcomponents of the total cross section [104, 105].

The photon–proton total cross section is related to the observed electron–proton cross section via the relation

$$\frac{d^2\sigma_{ep}}{dy\,dQ^2} = \frac{\alpha}{2\pi Q^2}\left(\frac{1+(1-y)^2}{y} - \frac{2(1-y)}{y}\frac{Q^2_{min}}{Q^2}\right)\sigma^{tot}_{\gamma^*p}(W_{\gamma^*p}), \quad (3.1.1)$$

where y is the fraction of the electron beam energy carried by the interacting quasi-real photon. The parameter Q^2_{min} is the minimum virtuality kinematically allowed by the photon energy and the electron mass (see (1.4.8)). Reconstruction algorithms for the photon virtuality Q^2 and the center-of-mass energy W_{γ^*p} employed by both the H1 and the ZEUS collaboration exploit the information from their luminosity-monitoring small-angle electron detectors, which measure the final-state angle θ and energy E'_e of the scattered electron:

$$y \approx \frac{E_\gamma}{E_e} = 1 - \frac{E'_e}{E_e}, \quad (3.1.2)$$

$$Q^2 \approx 4E_eE'_e\sin^2\theta_e/2, \quad (3.1.3)$$

$$W_{\gamma^*p} \approx \sqrt{4E_\gamma E_p}. \quad (3.1.4)$$

The photon and electron energies are expressed in the laboratory reference system and a small-scattering-angle approximation has been made. The acceptance of the electron detectors is limited to final-state electron angles which deviate from the electron beam flight direction by less than 5 mrad, resulting in measured Q^2 ranges $Q^2_{min} < Q^2 < 0.01$ GeV^2, with an average of about 5×10^{-4} GeV^2. The H1 experimental result employed photon energies ranging between 8.3 GeV and 19.3 GeV during dedicated running in 1994, covering a much wider range than that of the ZEUS analysis, which used data taken in 1992 and limited the photon energy to a region where the efficiencies were well understood: $8.5 < E_\gamma < 11.5$ GeV. The results for the total cross section were

$$\sigma^{tot}_{\gamma^*p} = 165 \pm 2\text{ (stat)} \pm 11\text{ (sys) }\mu\text{b for }\langle W_{\gamma^*p}\rangle = 200\text{ GeV (H1)},$$

$$\sigma^{tot}_{\gamma^*p} = 143 \pm 4\text{ (stat)} \pm 17\text{ (sys) }\mu\text{b for }\langle W_{\gamma^*p}\rangle = 180\text{ GeV (ZEUS)}.$$

Comparisons with earlier photoproduction measurements confirmed that the weak energy dependence (a mere 30% increase as W_{γ^*p} increases from 20 to 200 GeV), characteristic of the soft diffractive processes observed for hadronic cross sections, applies to photoproduction as well.

We turn now to measurements of the total photon–proton cross section with virtual photons at HERA. Such a cross section is well defined in a kinematic region where the lifetime of the virtual photon exceeds the interaction time associated with the longitudinal extent of the proton in its rest frame. This region is characterized by squared photon momenta which greatly exceed their virtuality, precisely the region measured at HERA. If we designate k_γ as the momentum of the virtual photon in the rest frame of the proton, then we can estimate its lifetime by applying the Heisenberg uncertainty relation to the amount of energy by which the photon is off-shell as follows [72, 106, 107]:

$$\Delta t \geq \frac{1}{2\Delta E}, \tag{3.1.5}$$

with

$$\Delta E = k_\gamma - \sqrt{k_\gamma^2 - Q^2}. \tag{3.1.6}$$

The HERA results we discuss cover a kinematic range

$$10^{-1} < Q^2 < 5 \times 10^3 \text{ GeV}^2, \quad 4 \times 10^2 < k_\gamma < 4 \times 10^4 \text{ GeV}. \tag{3.1.7}$$

For the kinematic region in which k_γ^2 dominates Q^2, we can approximate ΔE as follows:

$$\Delta E \approx \frac{1}{2} \frac{Q^2}{k_\gamma}. \tag{3.1.8}$$

Since the Lorentz-invariant variable ν is the photon energy in the proton rest frame we can write to a good approximation

$$\Delta E \approx \frac{1}{2} \frac{Q^2}{\nu} = m_{\mathrm{p}} x. \tag{3.1.9}$$

The speed of the virtual photon in the proton rest frame, v_γ, is only slightly greater than the vacuum speed of light:

$$v_\gamma = \frac{k_\gamma}{\sqrt{k_\gamma^2 - Q^2}} \approx 1 + \frac{1}{2} \frac{Q^2}{k_\gamma^2}, \tag{3.1.10}$$

so the condition for a well-defined $\gamma^* \mathrm{p}$ cross section becomes

$$\frac{1}{2m_{\mathrm{p}} x} > 2R_{\mathrm{p}}, \tag{3.1.11}$$

where R_{p} is the radius of the proton. Letting $R_{\mathrm{p}} = 0.8$ fm, we obtain the condition

$$\boxed{x < 0.06.} \tag{3.1.12}$$

In order to relate this condition to the center-of-mass energy we recall

$$W^2 = Q^2 \left[\frac{1}{x} - 1 \right] + m_{\mathrm{p}}^2 \approx \frac{Q^2}{x} \tag{3.1.13}$$

and observe that our condition becomes

$$\boxed{\frac{W^2}{Q^2} > 17.}$$ (3.1.14)

In order to relate the total γ^*p cross section to the differential deep inelastic electron–proton cross section we require a convention for the virtual photon flux K.[1] The convention of Hand [108],

$$K = \nu - \frac{Q^2}{2\,m_{\mathrm{p}}} = \frac{Q^2(1-x)}{2m_{\mathrm{p}}x},$$ (3.1.15)

is defined such that the photon–proton center-of-mass energy,

$$W^2 = m_{\mathrm{p}}^2 + 2m_{\mathrm{p}}K,$$ (3.1.16)

corresponds to a specific value for the flux independent of Q^2, which is convenient for the comparison of photoproduction with real and virtual photons. We can now relate the differential electron–proton cross section to the photoproduction cross section [106, 109, 73]:

$$\frac{\mathrm{d}^2\sigma^{\mathrm{ep}}}{\mathrm{d}x\mathrm{d}Q^2} = \Gamma_{\mathrm{T}}(x, Q^2)\,(\sigma_{\gamma^*\mathrm{p}}^{\mathrm{T}} + \varepsilon\,\sigma_{\gamma^*\mathrm{p}}^{\mathrm{L}}),$$ (3.1.17)

with the polarization parameter ε defined as

$$\varepsilon = \frac{2(1-y)}{1 + (1-y)^2 - 2(1-y)Q_{\mathrm{min}}^2/Q^2},$$ (3.1.18)

and where

$$\Gamma_{\mathrm{T}}(x, Q^2) = \frac{\alpha}{\pi}\frac{m_{\mathrm{p}}K}{Q^4}\left[1 + (1-y)^2 - 2(1-y)Q_{\mathrm{min}}^2/Q^2\right].$$ (3.1.19)

In the approximation $Q^2 \gg Q_{\mathrm{min}}^2$ the polarization parameter reduces to

$$\varepsilon = \frac{1-y}{1-y+y^2/2},$$ (3.1.20)

which deviates little from unity in the kinematic region covered by the HERA data, resulting in equal sensitivity to the longitudinal and transverse cross sections.

Following (1.4.18) we can write the photon–proton cross sections in terms of the structure functions F_1 and F_2, neglecting the contribution from F_3, which is significant only at higher values of Q^2, but including terms which contribute at low Q^2 and high x:

---

[1] A flux of virtual particles is an ill-defined concept in quantum field theory; only the product of the flux with the cross section is unambiguously defined. The conventions for defining this flux share a common limiting value as $Q^2 \to 0$ with fixed photon energy, which is the flux as defined for real photons.

$$\sigma_{\gamma^*p}^{T} = \frac{4\pi^2\alpha}{Q^2(1-x)}\, 2xF_1(x,Q^2),\tag{3.1.21}$$

$$\sigma_{\gamma^*p}^{L} = \frac{4\pi^2\alpha}{Q^2(1-x)}\left[\frac{Q^2+4\,m_p^2 x^2}{Q^2}\,F_2(x,Q^2) - 2xF_1(x,Q^2)\right].\tag{3.1.22}$$

In the approximation $x^2 m_p^2 \ll Q^2$,

$$\sigma_{\gamma^*p}^{L} \approx \frac{4\pi^2\alpha}{Q^2}F_{L}(x,Q^2),\tag{3.1.23}$$

and the sum of the longitudinal and transverse cross sections is simply

$$\sigma_{\gamma^*p}^{T} + \sigma_{\gamma^*p}^{L} \approx \frac{4\pi^2\alpha}{Q^2(1-x)}F_2(x,Q^2).\tag{3.1.24}$$

This Q^2 dependence is shown in Fig. 3.1, which includes measurements by the ZEUS [29] and H1 [35] collaborations obtained with the data recorded in 1994. The magnitude of the cross section indeed scales with Q^{-2} in this region, as the Q^2 dependence of $F_2(x, Q^2)$ is weak, and the energy dependence is also independent of Q^2 over a wide range in Q^2. The energy dependence is seen to be strong, the cross section rising by more than a factor of two as the energy increases from 50 to 200 GeV. Given the simple relationship between W_{γ^*p}, x, and Q^2 in this kinematic region (see (3.1.13)), this rise is easily identified as equivalent to the strong x dependence of the structure function F_2 discussed in Sect. 1.4.2.

Having observed the strong energy dependence of the proton interaction cross section for virtual photons when compared to that of the cross section for real photons, it becomes interesting to study the transition region. In a first attempt to investigate this region, the ZEUS collaboration used two experimental tricks in an early analysis of the data recorded in 1994. The first was to take advantage of the initial-state radiation from the electron to measure at effectively lower electron energy and hence lower Q^2. The second was to take advantage of HERA accelerator physicists' ability to manipulate the electron RF structure such that the interaction vertex was moved 67 cm in the forward direction, increasing the acceptance for electrons scattered at small angles. In this manner it became possible to measure the virtual-photon–proton cross section down to Q^2 values of 1.5 GeV^2 [110], as shown in Fig. 3.2. In this plot we compare as well the very different energy dependences of the photoproduction and deep inelastic cross sections. The solid line indicates the expectations based on phenomenological Regge theory [114], which describe the photoproduction measurements well but are inconsistent with the virtual-photon–proton measurements. The dashed line corresponds to the limit of validity of the cross section definition specified by (3.1.12), the cross section being well defined at high W. The measurements at higher values of x [30, 111, 112] exhibit a threshold-type behavior resulting from the short photon lifetime compared to the interaction time associated with the longitudinal size of the proton in its rest frame. The conclusion from these measurements at low x and low Q^2 is that the strong energy dependence holds down to Q^2 values of 1.5 GeV^2.

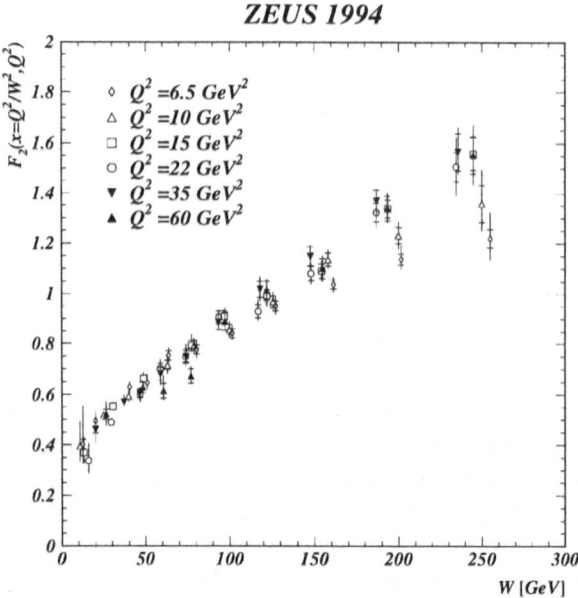

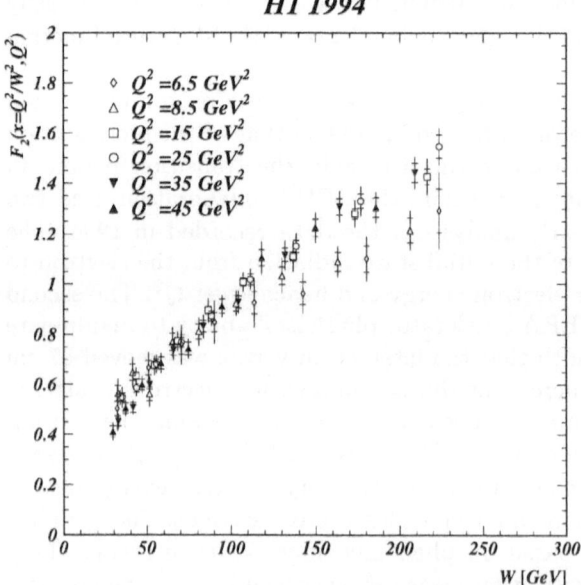

Fig. 3.1. The value of $F_2(x = Q^2/W_{\gamma^*p}^2, Q^2)$ as a function of the γ^*p center-of-mass energy W_{γ^*p} for various values of Q^2 as measured by the ZEUS [29] and H1 [35] collaborations with data recorded in 1994. The inner error bars represent the statistical uncertainties and the outer error bars show the quadratic sum of statistical and systematic uncertainties

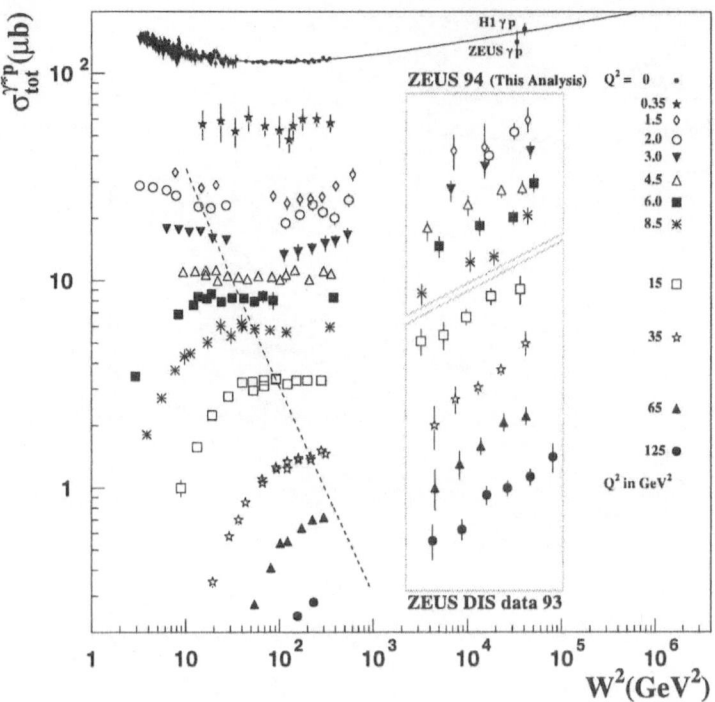

Fig. 3.2. The value of the total photon–proton cross section $\sigma_{\gamma^*p}^{tot}$ as a function of the squared center-of-mass energy for various values of Q^2 as measured by the ZEUS collaboration with subsamples (see text for details) of the 1994 data [110], compared to measurements at higher Q^2 using the 1993 data [28], the HERA photoproduction cross section measurements [104, 105], and the low-energy deep inelastic [30, 111, 112] and photoproduction [113] measurements. The curve represents a parametrization of the photoproduction measurements based on phenomenological Regge theory [114]. The dashed line connects points where x=0.1

The HERA measurements employing the full data set from the 1994 data-taking period [29, 35] are shown in Fig. 3.3 and compared to calculations using the ALLM proton structure function parametrizations [115], which exhibit consistency with the data over the entire range in Q^2. Substantial gains in the kinematic range were achieved, and the region covered by these measurements now reaches that of the lower-energy deep inelastic fixed-target experiments. The transition region from the threshold behavior at high x to the high W region covered by the HERA data is now precisely measured. The slope in the energy dependence is seen to decrease with decreasing Q^2, but these measurements also do not permit a judgement as to the Q^2 value for which the transition to the Regge behavior occurs.

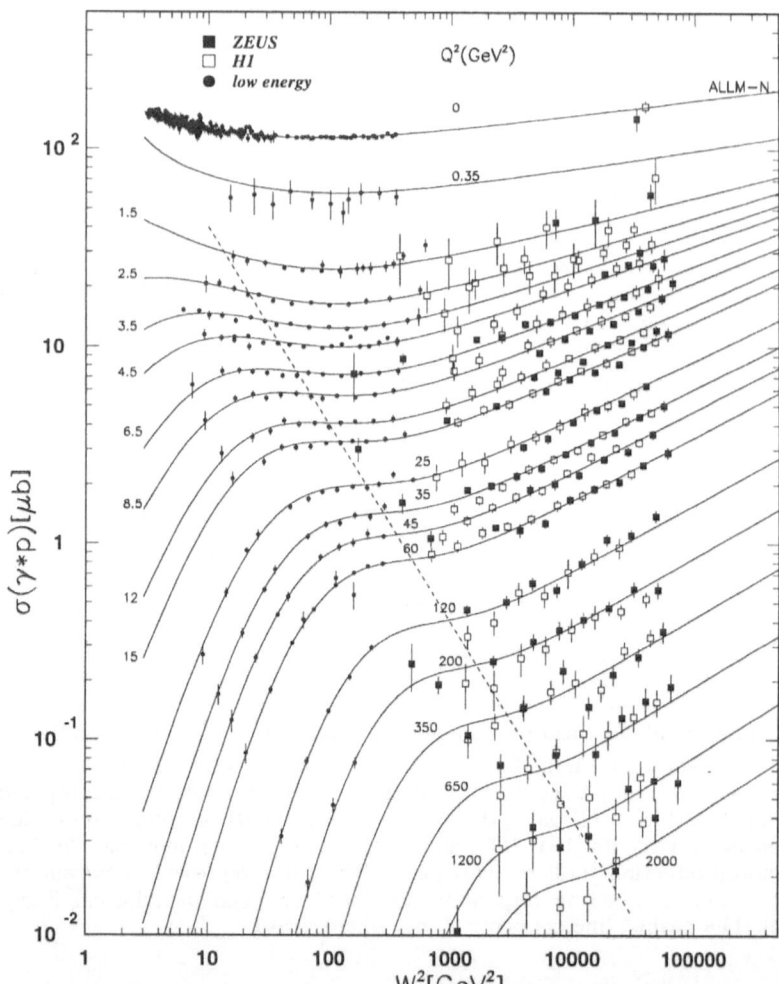

Fig. 3.3. The value of the total photon–proton cross section $\sigma_{\gamma^*p}^{tot}$ as a function of the squared center-of-mass energy for various values of Q^2 as measured by the ZEUS [29] and H1 [35] collaborations with the 1994 data and by the NMC collaboration [30] at CERN. The curves represent calculations using the ALLM proton structure function parametrizations [115]. The dashed line connects points where $x=0.1$

The beam-pipe calorimeter installed by the ZEUS collaboration prior to the 1995 data-taking period has permitted a more detailed investigation of the low-Q^2 region. Preliminary measurements [116] of the total virtual-photon-proton cross sections for $0.16 < Q^2 < 0.57$ GeV^2 and $130 < W < 230$ GeV are shown in Fig. 3.4, along with the ZEUS results from the 1994 data at

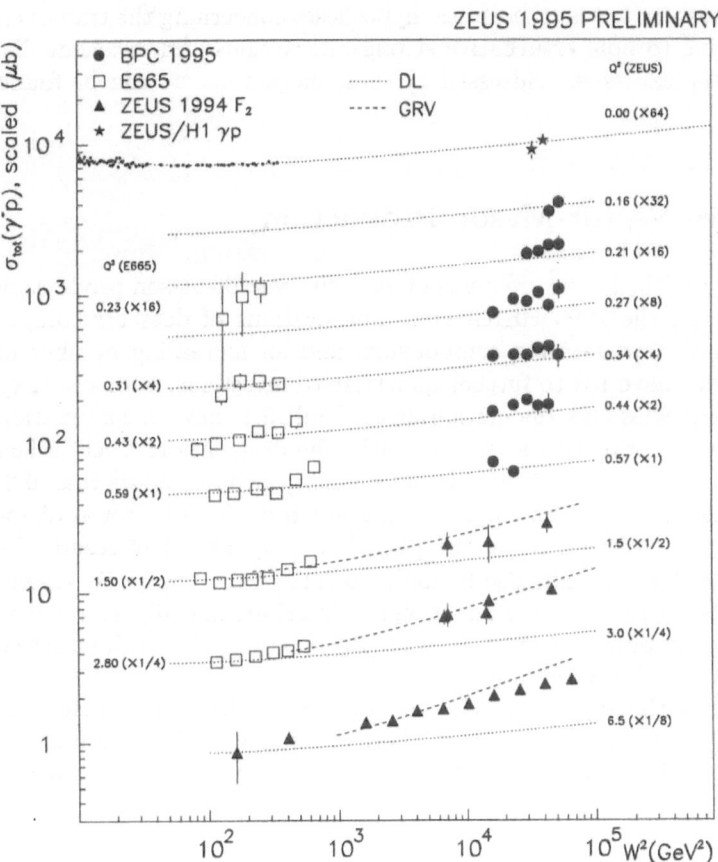

Fig. 3.4. The value of the total photon–proton cross section $\sigma_{\gamma^*p}^{tot}$ as a function of the squared center-of-mass energy for various values of Q^2 as measured by the ZEUS collaboration [116] with the 1994 and 1995 data compared to results from the E665 collaboration [112] at FNAL. The low-energy photoproduction measurements [113] are also shown. Calculations based on phenomenological Regge theory [114] are indicated as dotted lines and perturbative QCD calculations [117] are shown as dashed lines. This comparison of the HERA data with the fixed-target muon scattering measurements at higher values of x indicates that the transition region in Q^2 between the two types of energy dependence lies in the region $0.16 < Q^2 < 3.0$ GeV^2

low Q^2. The comparison to data from the E665 collaboration [112] at higher x using extrapolations based on Regge-type energy dependence [114] and that expected from QCD calculations [117] indicates that the transition occurs between Q^2 values of 0.16 and 3.0 GeV^2. While this conclusion is mitigated by the necessity of comparing results from two different experiments and by the 5% normalization uncertainty in the ZEUS results not shown in Fig. 3.4, it appears that future measurements in this kinematic region will provide the information necessary to test theoretical hypotheses concerning the transition from perturbative to nonperturbative strong interactions. An excellent discussion of the topical issues addressed by these measurements can be found in [118].

3.2 Exclusive Vector-Meson Production

Since the first published measurements of exclusive vector-meson production appeared following the 1993 HERA run, ameliorations of detector components, addition of new detector components, and an increasing number of analysis strategies have led to further quantitative investigation of a variety of final-state topologies and kinematic ranges. Table 3.1 shows a compilation of results published prior to the end of 1996. The designation "Inel" refers to processes which result in a value for the elasticity variable z less than 0.8. Results which include quasi-elastic scattering accompanied by proton dissociation ($z \lesssim 1$) are referred to as "El/pDiss". A comparison of results for ρ^0, ϕ, and J/ψ at high Q^2 can also be found in [119]. Preliminary H1 results from the 1995 data run for J/ψ production are described in [120]. The review of HERA results on diffractive processes in [121] also includes an instructive chapter on vector meson production.

We now discuss the experimental issues common to the measurements for each type of vector meson investigated. Thereafter follows a discussion of the considerations particular to each of these vector mesons and to each decay mode studied.

3.2.1 General Experimental Considerations

The Photon Lifetime. The discussion of the validity of the photon–proton cross section definition follows closely that in Sect. 3.1, with the replacement

$$Q^2 \rightarrow Q^2 + M_V^2, \tag{3.2.25}$$

where M_V is the mass of the final-state vector meson. Thus the limiting value of x is lower than the value of 0.06 found in (3.1.12), but all of the measurements we discuss in the following cover kinematic ranges in which the lifetime of the photon in the proton rest frame far exceeds the interaction time associated with the longitudinal extent of the proton in its rest frame.

Table 3.1. Measurements of single neutral vector-meson production at HERA published by the H1 and ZEUS collaborations prior to the end of 1996. The event statistics presented are either quoted directly or estimated from information provided in the respective reference. They are corrected for nonresonant backgrounds which appear in the invariant mass distributions, but not for the backgrounds from other sources, which are shown in the column "Bckgd". The transverse momentum ranges shown refer to the vector momentum sum of particles observed in the central detectors and approximate the ranges of momentum transfer at the proton vertex. Information which was unavailable is designated by "$n.d.$"

VM	EI/Inel/pDiss	Decay channel	Q^2 range (GeV²)	$\langle Q^2\rangle$ (GeV²)	W range (GeV)	$\langle W\rangle$ (GeV)	p_T^2 range (GeV²)	Events	Bckgd (%)	Ref
ρ^0	EI	$\pi^+\pi^-$	10^{-10}–4	10^{-1}	60–80	70	0–0.5	6381	15±6	[85]
ρ^0	EI	$\pi^+\pi^-$	10^{-10}–0.5	10^{-1}	40–80	55	0–0.5	358	33±10	[122]
ρ^0	EI	$\pi^+\pi^-$	10^{-8}–0.01	10^{-3}	164–212	187	0–0.5	1000	31±8	[122]
ρ^0	EI	$\pi^+\pi^-$	7–25	11	42–134	79	0–0.6	82	26±18	[123]
ρ^0	EI	$\pi^+\pi^-$	8–50	13	40–140	81	0–0.5	160	10±8	[124]
ρ^0	EI	$\pi^+\pi^-$	10^{-10}–1	10^{-1}	50–100	73	0.07–0.40	1653	7±1	[19]
ρ^0	EI/pDiss	$\pi^+\pi^-$	7–36	11	60–180	120	0–0.8	260	11±8	[125]
ρ^0	EI	$\pi^+\pi^-$	10^{-10}–4	10^{-1}	50–100	70	0–0.5	80 000	22±3	[126]
ρ^0	EI/pDiss	$\pi^+\pi^-$	0.25–0.85	0.5	20–90	55	0–0.6	1500	27±9	[127]
ρ^0	EI/pDiss	$\pi^+\pi^-$	7–36	11	60–180	120	0–0.8	90	18±6	[125]
ρ^0	EI/pDiss	$\pi^+\pi^-$	10^{-9}–0.01	10^{-3}	85–105	95	0.5–4.0	3000	$n.d.$	[128]
ρ'	EI	$\pi^+\pi^-\pi^+\pi^-$	4–50	7	40–140	85	0–0.6	70	9±6	[129]
ω	EI	$\pi^+\pi^-\pi^0$	10^{-9}–4	10^{-1}	70–90	80	0–0.6	170	16±9	[130]
ϕ	EI	K^+K^-	10^{-10}–4	10^{-1}	60–80	70	0.1–0.5	570	25±10	[131]
ϕ	EI	K^+K^-	7–25	11	42–134	95	0–0.6	37	22±17	[132]
ϕ	EI	K^+K^-	6–20	10	42–134	88	0–1.0	32	9±8	[133]
J/ψ	EI	e^+e^-,μ^+,μ^-	10^{-10}–4	10^{-1}	40–140	95	0–1.0	104	17±13	[134]
J/ψ	EI	e^+e^-,μ^+,μ^-	10^{-10}–4	10^{-1}	40–140	90	0–1.0	620	15±7	[135]
J/ψ	Inel	e^+e^-	10^{-9}–4	10^{-1}	110–160	135	0–1.0	25	$n.d.$	[135]
J/ψ	Inel	$\mu^+\mu^-$	10^{-10}–4	10^{-1}	60–130	95	0–1.0	65	$n.d.$	[135]
J/ψ	EI	$e^+e^-,\mu^+\mu^-$	10^{-10}–4	10^{-1}	30–150	95	0–2.0	430	12±11	[136]
J/ψ	EI/pDiss	$e^+e^-,\mu^+\mu^-$	10^{-10}–4	10^{-1}	30–150	105	0–2.0	320	$n.d.$	[136]
J/ψ	Inel	$\mu^+\mu^-$	10^{-10}–4	10^{-1}	30–150	105	0–5.0	85	3	[136]
J/ψ	EI	$e^+e^-,\mu^+\mu^-$	8–40	16	30–150	95	$n.d.$	25	25±11	[124]
J/ψ	EI	$e^+e^-,\mu^+\mu^-$	7–25	11	42–134	80	$n.d.$	14	15±7	[119]
$\psi(2S)$	EI	$J/\psi\,\pi^+\pi^-$	10^{-10}–4	10^{-1}	40–160	80	$n.d.$	25	$n.d.$	[137]

The Photon Flux. The effective photon flux is obtained from the relationship between the differential electron-proton cross section and the longitudinal and transverse photon–proton cross sections $\sigma^{L}_{\gamma^{*}p\to Vp}$ and $\sigma^{T}_{\gamma^{*}p\to Vp}$ as follows:

$$\frac{d^2\sigma_{ep\to eVp}}{dy\,dQ^2} = \frac{\alpha}{2\pi y Q^2}\left[\left(1+(1-y)^2-2(1-y)\frac{Q^2_{min}}{Q^2}\right)\sigma^{T}_{\gamma^{*}p\to Vp}\right.$$

$$\left. +\,2\,(1-y)\,\sigma^{L}_{\gamma^{*}p\to Vp}\right], \qquad (3.2.26)$$

where $Q^2_{min} = m^2_e y^2/(1-y)$. For photoproduction studies the ratio of the two partial cross sections can be estimated using VMD assumptions [138]:

$$\frac{\sigma^{L}_{\gamma^{*}p\to Vp}(W_{\gamma^{*}p},Q^2)}{\sigma^{T}_{\gamma^{*}p\to Vp}(W_{\gamma^{*}p},Q^2)} = \xi\frac{Q^2}{M^2_V}, \qquad (3.2.27)$$

with ξ of order unity, and the transverse cross section is expected to scale with Q^2 as

$$\sigma^{T}_{\gamma^{*}p\to Vp}(W_{\gamma^{*}p},Q^2) = \left(1+\frac{Q^2}{M^2_V}\right)^{-2}\sigma^{T}_{\gamma^{*}p\to Vp}(W_{\gamma^{*}p},Q^2=0). \quad (3.2.28)$$

The flux falls rapidly with y and Q^2, while the $\gamma^{*}p$ cross section itself depends only weakly on these variables. Thus we can write to a good approximation[2]

$$\sigma^{tot}_{\gamma^{*}p\to Vp} \equiv \sigma^{T}_{\gamma^{*}p\to Vp}+\varepsilon\sigma^{L}_{\gamma^{*}p\to Vp} = \frac{\sigma_{ep\to eVp}}{\Phi_{\gamma^{*}/e}}, \qquad (3.2.29)$$

where ε is the virtual photon polarization parameter, defined as in (3.1.18), and the flux of photons is calculated as

$$\Phi_{\gamma^{*}/e} = \int_{y_{min}}^{y_{max}} dy \int_{Q^2_{min}}^{Q^2_{max}} dQ^2\, f_{\gamma^{*}/e}(y,Q^2), \qquad (3.2.30)$$

$$f_{\gamma^{*}/e}(y,Q^2) = \frac{\alpha}{2\pi y Q^2}\frac{1+(1-y)^2-2(1-y)\left(\dfrac{Q^2_{min}}{Q^2}-\xi\dfrac{Q^2}{M^2_V}\right)}{\left(1+\dfrac{Q^2}{M^2_V}\right)^2},$$

$$(3.2.31)$$

with Q^2_{min} defined as above and the other limits defined by the acceptance and event selection cuts. The integrand falls by an order of magnitude over the

---

[2] The designation "tot" refers here to the sum of longitudinal and transverse contributions to the elastic cross section, in contrast to its use in (3.1.1) and (3.1.24), where it referred to the $\gamma^{*}p$ interaction cross section summed over all final states.

y range covered by the measurements. The dependence on the vector-meson mass arises primarily from the longitudinal component and influences the flux calculation at a level of less than a few percent for the photoproduction studies. We note further that in the kinematic range of the HERA measurements the polarization parameter exceeds 0.97, resulting in roughly equal sensitivity to the transverse and longitudinal cross sections. The latter is negligibly small in the photoproduction studies, while at high Q^2 the two cross sections are of comparable magnitude. For $M_V^2 \gg Q^2 \gg m_e^2$ the relationship between the photon–proton cross section and the differential electron–proton cross section reduces to

$$\sigma_{\gamma^* p \to V p}^{tot} = \frac{1}{f_{\gamma^*/e}^T(y, Q^2)} \frac{d^2 \sigma_{ep \to eVp}}{dy \, dQ^2}, \tag{3.2.32}$$

where

$$f_{\gamma^*/e}^T(y, Q^2) = \frac{\alpha \left[1 + (1-y)^2\right]}{2\pi y Q^2}. \tag{3.2.33}$$

The assumption of weak energy dependence applied in the total cross section calculations as described above is invalid for the studies at high Q^2 and for the studies of J/ψ photoproduction. For such investigations the flux factor was calculated on an event-by-event basis.

Kinematic Reconstruction. The vector-meson event samples all required the reconstruction of the invariant mass of the final-state meson. With the sole exception of the π^0 in $\omega \to \pi^+ \pi^- \pi^0$ decay all decay products were identified as such via charged-particle tracking, which provided high-accuracy determination of their momentum three-vectors at the production vertex. Using ad hoc mass assumptions for these tracks, the vector mesons were identified via peaks in invariant mass spectra, and their momentum four-vectors were determined. In the photoproduction samples no other information was available, and the $\gamma^* p$ center-of-mass energy was determined via the relation

$$W_{\gamma^* p} = 2E_p (E - P_z)_V, \tag{3.2.34}$$

where E_p is the initial-state proton energy and $(E - P_z)_V$ is the difference between the energy and the longitudinal momentum component of the vector meson. The accuracy provided by the tracking allowed a resolution of about 2 GeV in $W_{\gamma^* p}$. For the investigations at high Q^2 additional information on the scattered electron was available via tracking or calorimetric measurements. In this case the reconstruction strategies of the two experiments differed, as H1 employed the double-angle method [139] used in the structure function analyses and ZEUS used a method which exploited the assumption of exclusivity. Under this assumption, one can determine the energy of the scattered electron from its scattering angle θ_e and the vector-meson four-vector as follows:

$$E_e' = \frac{2E_e - (E - P_z)_V}{1 + \cos \theta_e}. \tag{3.2.35}$$

The resolution in the electron energy obtained in this manner was more than a factor of five better than that obtained from the direct calorimetric measurement.

A reasonable approximation to the momentum transfer t at the proton vertex (see (1.4.14)) is the negative squared total transverse momentum observed in the detector, i.e. that of the vector meson for photoproduction samples and

$$t = -p_T^2 = -\left[(p_x^e + p_x^V)^2 + (p_y^e + p_y^V)^2\right] \tag{3.2.36}$$

for the high-Q^2 samples, using either the double-angle electron momentum or the constrained electron momentum. For photoproduction studies in which the final-state electron was detected in one of the small-angle electron taggers, the momentum transfer was well determined by the transverse momentum of the vector meson alone $(Q^2 < 0.01\,\text{GeV}^2)$, while in the untagged photoproduction studies the smearing due to the unknown electron transverse momentum was corrected for via unfolding algorithms, and clean samples of high-t events were not obtained. The sole exception to these means of extracting the momentum transfer was the analysis exploiting data from the ZEUS Leading Proton Spectrometer, where a direct measurement of the scattered proton allowed a resolution in t of 0.01 GeV^2, dominated by the proton beam divergence.

Luminosity Measurement. The H1 and ZEUS experiments each base their determination of the luminosity used in the extraction of the electron–proton cross sections on the rate from the Bethe–Heitler bremsstrahlung process ep $\rightarrow$ e$'\gamma$p. This rate can be calculated to an accuracy of 0.2% in an approximation which ignores the proton structure and its recoil momentum [140]. The signature for the process is simple and clean, since the energies of the final-state electron and photon are measured in small-angle calorimeters and their sum equals the energy of the initial-state electron. Primary contributions to the uncertainty in the luminosity measurement are the resolution and acceptances of the calorimeters, and the corrections for contributions from bremsstrahlung on the residual gas in the beam pipe. This uncertainty improved from approximately 4% in the 1993 data to 2% in 1994. In none of the investigations described in this review does the uncertainty in the luminosity dominate the normalization uncertainty in the cross section. The primary contributions to the normalization uncertainty are discussed for each of the studies individually in Sect. 3.2.2.

Background Estimates. The primary background in all of the studies of exclusive vector-meson production save one has been from processes where the proton dissociates, since the fiducial coverage in the forward direction in each experiment does not provide a good measurement of the dissociative final state, being used primarily as a veto in the trigger and offline selection cuts. These background estimates ranged between 10% and 40%, depending on the exclusive process under study. Indeed the *uncertainty* in the background estimations, which were based on simulations of poorly-known physics processes, often dominated the uncertainty in the measured cross sections. The exception was the measurement which employed the ZEUS LPS to measure the final-state proton trajectory [19], allowing the reduction of background contamination from this source to a level of tenths of a percent.

3.2.2 Experimental Issues Particular to the Decay Channels Investigated

$\rho^0 \rightarrow \pi^+\pi^-$. This process was the first exclusive process to be studied at HERA [85], owing to its high rate and clean signature in the photoproduction data sample. Trigger conditions requiring an energy deposit of 0.5 GeV in the calorimeter, at least one track in the central tracking chamber, and a veto on energy deposits in the forward direction sufficed to reduce the trigger rate to an acceptable level. The trigger efficiency was about 40% and the geometrical acceptance 7%. Figure 3.5 shows the mass spectrum

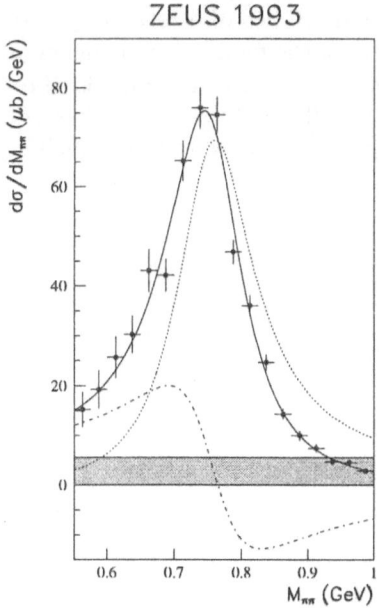

Fig. 3.5. The differential elastic $\pi^+\pi^-$ production cross section $d\sigma/dM_{\pi\pi}$ for the photoproduction study by the ZEUS collaboration using data recorded in 1993. The curves indicate the results of a fit to the data with contributions from a relativistic Breit–Wigner resonant term interfering with a nonresonant background term. The dotted line shows the Breit–Wigner term, the dash-dotted line the interference term, and the shaded band indicates the statistical uncertainty in the determination of the background term. Only statistical errors are shown. The horizontal error bars indicate the size of the bins

obtained. The sum of a relativistic Breit–Wigner resonance term, a nonresonant dipion background term, and an interference term [50, 141, 142, 143] was used to fit this distribution. Further models for this line shape [144] are discussed in the review of Bauer et al. [49]. This measurement covered the range $60 < W_{\gamma^* p} < 80$ GeV and $|t| < 0.5$ GeV^2. The background contamination was estimated to be about 15%, dominated by the contribution from proton dissociative processes. The systematic uncertainty in the total elastic cross section determination was 17%, with the primary contribution coming from the uncertainty in the acceptance calculation.

The H1 collaboration performed a similar study with the 1993 data and supplemented it with data from a dedicated run during the 1994 data-taking period, when the interaction vertex was shifted 67 cm in the forward direction and the trigger was based on tagging the scattered electron in the luminosity monitor [122]. This method allows measurement of the exclusive cross section at values of $W_{\gamma^* p}$ similar to those of the total cross section measurements: $164 < W_{\gamma^* p} < 212$ GeV. The background of $(31 \pm 8)\%$ was dominated by dissociative processes and the systematic error in the cross section measurement of 18% arose primarily from the uncertainty in the background subtraction.

In 1994 the ZEUS LPS began operation and studies of elastic ρ^0 photoproduction with a direct, accurate measurement of the momentum transfer at the proton vertex were thus made possible [19]. The acceptance of the LPS restricted the data to the kinematic region $0.073 < |t| < 0.40$ GeV^2. A fit to the invariant mass spectrum as in the earlier ZEUS study is shown in Fig. 3.6. The

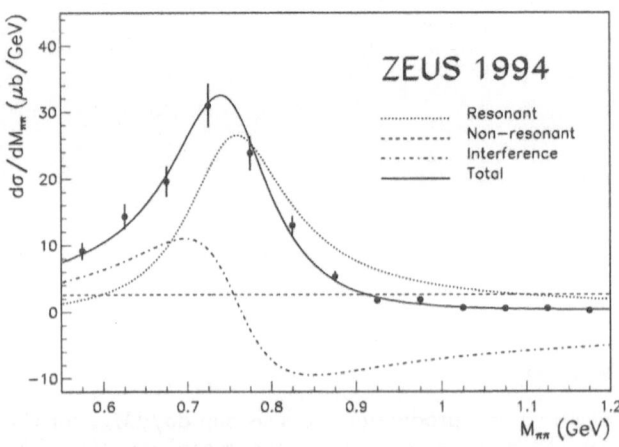

Fig. 3.6. The differential elastic $\pi^+\pi^-$ photoproduction $d\sigma/dM_{\pi\pi}$ restricted to the kinematic region $0.073 < |t| < 0.40$ GeV^2. This region was defined by the acceptance of the ZEUS Leading Proton Spectrometer [19], which provided a direct measurement of the final-state proton momentum. The solid curve indicates the results of a fit to the data with contributions from a relativistic Breit–Wigner resonant term, a nonresonant background term, and an interference term, which are shown by the broken lines. Only statistical errors are shown

requirement of a final-state proton in the LPS reduced the contribution from proton dissociative processes to less than 0.5%. The dominant background came from events with a ρ^0 in the main detector in accidental coincidence with an LPS track from a charged particle produced in upstream beam–gas interactions, and was estimated to be $(5.0 \pm 0.6)\%$. The dominant contribution to the systematic uncertainty of 12% in the cross section measurement was the uncertainty in the trigger efficiency.

Preliminary results on ρ^0 production from the 1994 data have been presented by the ZEUS collaboration [126], featuring an order-of-magnitude increase in statistics in comparison to the earlier publication. Figure 3.7 shows the invariant mass spectra obtained in this study for various ranges of the ρ^0

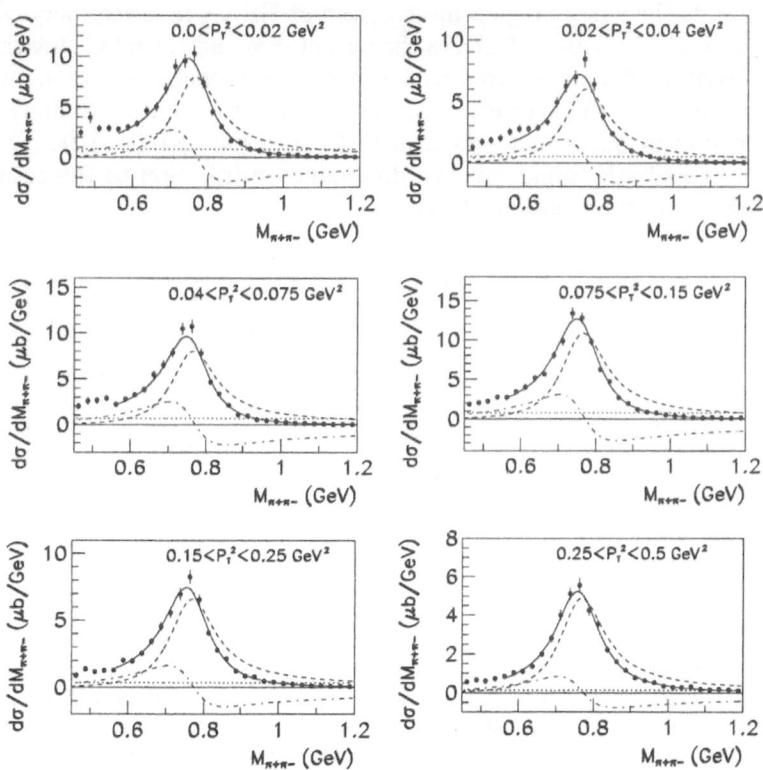

Fig. 3.7. The differential elastic $\pi^+\pi^-$ production cross section $d\sigma/dM_{\pi\pi}$ for the photoproduction study by the ZEUS collaboration using data recorded in 1994. The solid curve indicates the results of a fit to the data with contributions from a relativistic Breit–Wigner resonant term, a nonresonant background term, and an interference term, which are shown by the broken lines. Only statistical errors are shown

transverse momentum. The statistical power of this new study shows clearly that the relative contribution from the nonresonant dipion background is reduced as the ρ^0 transverse momentum increases. A subsample of this data set for which LPS information was available served to extract an estimate of the proton dissociative background of $(22 \pm 3)\%$. The total systematic uncertainty in the cross section determination was about 10%.

The ZEUS collaboration has presented preliminary results from the 1995 data-taking period in the low-Q^2 region $0.25 < Q^2 < 0.85$ GeV^2 [127]. The Q^2 acceptance was determined by the size of the beam-pipe calorimeter used to provide a trigger and to measure the positron scattering angle between 17 and 35 mrad with a resolution of 0.3 mrad. Since the constrained method for determining the electron energy yields a resolution of 1% and Q^2 depends on the square of the scattering angle, this angular resolution largely determines the resolution in Q^2 of about 4% (see (1.4.3)). The BPC measurement extends the energy region investigated at HERA to lower energies: $20 < W_{\gamma^*p} < 90$ GeV. Figure 3.8 shows the acceptance and the $\pi^+\pi^-$ invariant mass spectrum, which was fitted with the sum of resonant, nonresonant, and interference terms. The primary source of background was estimated to arise from proton dissociative processes: $(27 \pm 6 \, (\text{stat}) \pm 6 \, (\text{sys}))\%$. The uncertainty in this background was the dominant contribution to the total systematic uncertainty estimate of 10%.

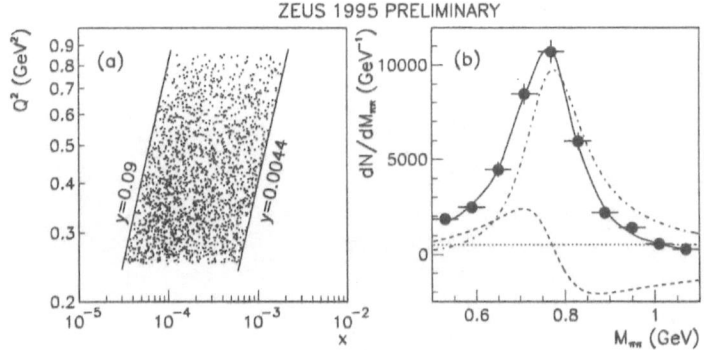

Fig. 3.8. (a) The kinematic region accepted and (b) the $\pi^+\pi^-$ invariant mass distribution for elastic ρ^0 production at intermediate Q^2 measured by the ZEUS collaboration, based on data recorded in 1995 using the newly installed beam-pipe calorimeter. The solid curve on the mass spectrum indicates the results of a fit to the data with contributions from a relativistic Breit–Wigner resonant term, a nonresonant background term, and an interference term, which are shown by the broken lines. Only statistical errors are shown

The H1 and ZEUS experiments have each published observations of elastic ρ^0 production at values of Q^2 near 10 GeV^2 [119, 123, 124, 125, 145, 146]. Trigger efficiencies and acceptances for these processes were high ($> 50\%$) owing to the high energy and angle of the scattered electron. The data samples

typically consisted of a few hundred events, which sufficed to show that the
mass spectra were more symmetric than those observed in photoproduction,
and Breit–Wigner distributions gave reasonable fit results, the widths consistent with the Particle Data Group value of 150 MeV. The backgrounds were
around 20%, dominated by proton dissociation. The uncertainties in the determination of the exclusive cross section were around 10%, arising from the
theoretical uncertainty in the dependence of the dissociative background rate
on the invariant mass of the dissociated final state.

$\rho' \rightarrow \pi^+\pi^-\pi^+\pi^-$. The H1 collaboration has reported preliminary results
on the observation of a resonance with mass $(1.57 \pm 0.02 \text{ (stat)} \pm 0.01 \text{ (sys)})$
GeV and width $(0.18 \pm 0.06 \text{ (stat)} \pm 0.07 \text{ (sys)})$ GeV in the $\pi^+\pi^-\pi^+\pi^-$ invariant mass spectrum for $40 < W_{\gamma^*p} < 140$ GeV and $Q^2 > 4$ GeV^2. A sample of 108 events was observed in the peak shown in Fig. 3.9, of which
about a third was attributed to nonresonant background. A peak at the
ρ^0 mass was observed in the spectrum of $\pi^+\pi^-$ masses, of which there are
two such combinations in each of the events, indicating a decay to $\rho^0\pi^+\pi^-$.
The production rate relative to that for the ρ^0 meson was measured to be
$(0.36 \pm 0.07(\text{stat}) \pm 0.11(\text{sys}))$. The authors offered speculation that this resonance arises via constructive interference of the $\rho^0[1450]$ and $\rho^0[1700]$ res-
onances. The uncertainty in the elastic cross section was about 30%, with
comparable contributions from statistics and from the uncertainty in the
nonresonant background subtraction.

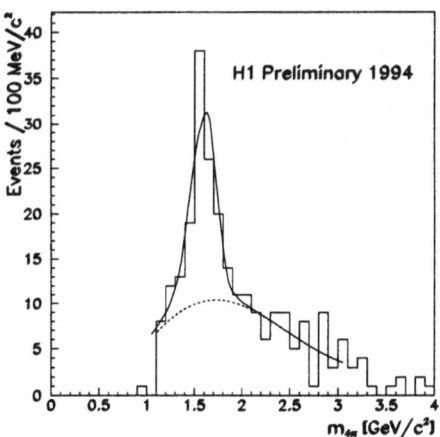

Fig. 3.9. The $\pi^+\pi^-\pi^+\pi^-$ invariant mass spectrum for the $Q^2 > 4$ GeV^2 data
sample in 1994 from the H1 collaboration. The solid curve shows the result of a fit
to the sum of a single resonance and a polynomial background which is shown by
the dashed line

$\omega \rightarrow \pi^+\pi^-\pi^0$. The ZEUS collaboration has reported a measurement of elastic ω photoproduction [130], where the π^0 decay photons were detected in an array of 3×3 cm^2 silicon diodes located at a depth of 3.3 X_0 inside the rear calorimeter. The average photon energy was 500 MeV with a broad spectrum of r.m.s. width about 200 MeV. The reconstructed $\gamma\gamma$-mass spectrum is shown in Fig. 3.10, as is the $\pi^+\pi^-\pi^0$ invariant mass spectrum showing the ω peak. The second peak in the three-pion invariant mass distribution was attributed to the decay $\phi \rightarrow \pi^+\pi^-\pi^0$. The photon energy resolution in the rear calorimeter determined the resolution in the γ^*p center-of-mass energy, which was reconstructed as in (3.2.34). The resolution in $W_{\gamma^*\mathrm{p}}$ was 6% and that in p_T^2 was 0.04 GeV^2. The mass resolution for the ω was about 30 MeV. The primary contribution to reconstruction inefficiency was that associated with the reconstruction of the π^0; the product of trigger efficiency and acceptance was about 1%. The background from proton dissociation was estimated to be (16 ± 9)%. The uncertainty in the total cross section of 20% was dominated by the systematic uncertainties arising from the acceptance calculation and the background estimation.

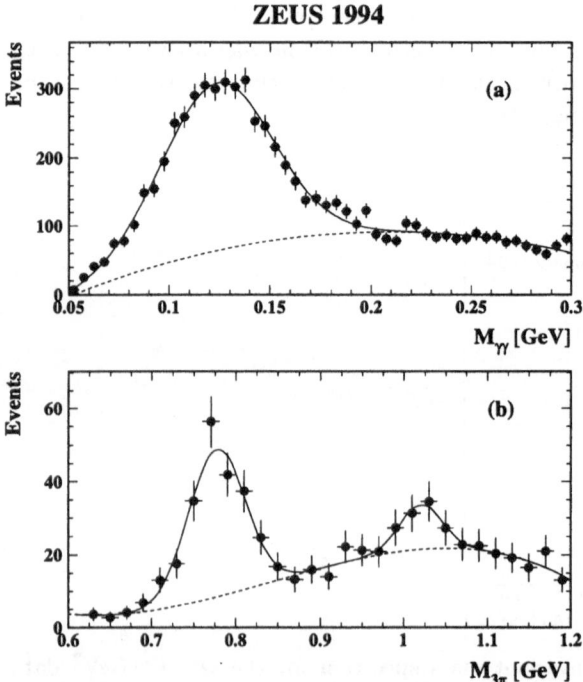

Fig. 3.10. Invariant mass distributions measured by the ZEUS collaboration for the process $\mathrm{e^+p} \rightarrow \mathrm{e^+p}\,\pi^+\pi^-\,\gamma\gamma$. **(a)** The two-photon invariant mass calculated using the photon positions as measured in the silicon array in the rear calorimeter of the ZEUS detector. **(b)** The three-pion invariant mass distribution exhibiting peaks corresponding to the decays of ω and ϕ mesons. In each plot the solid line indicates the results of the full fit used in the analysis and the dashed line shows the contribution by nonresonant backgrounds as determined by the fit procedure

$\phi \rightarrow K^+K^-$. An observation of elastic ϕ photoproduction has been reported by the ZEUS collaboration [131]. The triggering and event selection procedures for elastic ϕ production were very similar to those for the ρ^0 meson, since no πK separation was done. However, the 4.4 MeV natural width of the ϕ was comparable to the resolution in the invariant mass reconstruction using the two charged track trajectories. Therefore the ZEUS analysis fitted the mass spectrum with the sum of a Gaussian-smeared Breit–Wigner function and a background polynomial, as shown in Fig. 3.11. The background consisted primarily of ρ^0 decays where the kaon mass was mistakenly assigned to the decay products in the invariant mass calculation. The width of the Breit–Wigner function was fixed to the Particle Data Group value and the fit result for the Gaussian width was found to be 4 MeV, compatible with the expected experimental resolution. The detector acceptance for this process depended strongly on the total squared transverse momentum, so the kinematic range was restricted to $0.1 < |t| < 0.5$ GeV^2 and $60 < W_{\gamma^*p} < 80$ GeV, where the acceptance calculation was precise. This restriction in p_T^2 and the acceptance resulted in a proton dissociative background estimation of $(24 \pm 7 \text{ (stat)} \pm 6 \text{ (sys)})\%$, double that for the ρ^0. The uncertainty in this background was included in the statistical error estimation for the total cross section along with that from the determination of the ρ^0 background from the fit to the mass peak to yield a total statistical error of 20%. The systematic error includes roughly equal contributions from the acceptance determination, the trigger efficiency estimation, and from the extrapolation $t \rightarrow 0\,\text{GeV}^2$, totaling approximately 20%.

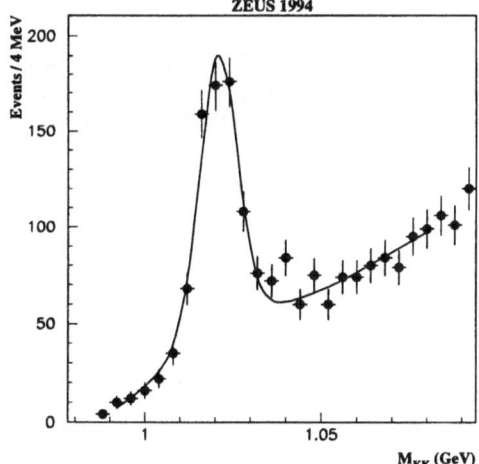

Fig. 3.11. The K^+K^- invariant mass spectrum for the photoproduction data sample in 1994 from the ZEUS collaboration. The solid curve shows the result of a fit to the sum of a single resonance and a background shape due primarily to ρ^0 decays

The high-Q^2 studies of exclusive ϕ production by the ZEUS [132] and
H1 [133] collaborations do not have the t-acceptance difficulty of the photoproduction study, enjoying $\approx$60% acceptance due to the ease of triggering on the
scattered electron. The resolution in the mass peak was similarly $\approx$4 MeV
(see Fig. 3.12), but the background from ρ^0 decays was smaller than for the
photoproduction results, totaling about 10%. The background from proton
dissociation was estimated to be $(22 \pm 8$ (stat) ± 15 (sys)%) in the ZEUS
study and (9 ± 8)% by the H1 collaboration. The statistical uncertainties in
the cross section determinations due to the small event samples were similar
for the two experiments at $\approx$20%, while the systematic uncertainty for H1
was estimated to be roughly half of the value of 30% quoted by ZEUS, owing
to more accurate estimates of the ρ^0 and proton dissociation backgrounds.

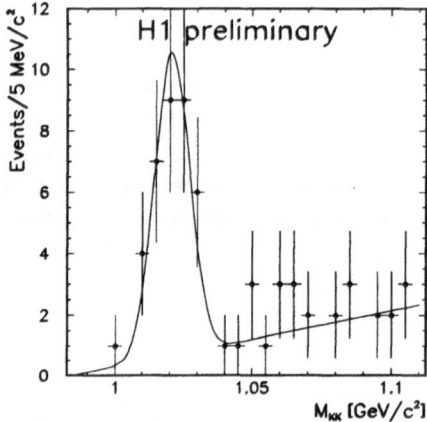

Fig. 3.12. The K^+K^- invariant mass spectrum for the 1994 data from the H1
collaboration for the range $6 < Q^2 < 20$ GeV^2. The solid curve shows the result of
a fit to a Gaussian-smeared Breit–Wigner distribution over a linear background due
primarily to ρ^0 decays

$J/\psi \rightarrow$ e$^+$e$^-$ **and** $\mu^+\mu^-$. Measurements of elastic and inelastic photoproduction of J/ψ mesons have been reported by both the ZEUS [134, 135]
and the H1 [120, 136, 147] collaborations. These studies were based on the
reconstruction of two tracks within the acceptance of the central detectors,
restricting them to the kinematic range $50 < W_{\gamma^*p} < 150$ GeV, with the exception of a sample based on the reconstruction of two electromagnetic clus-
ters in the rear calorimeter of the H1 detector which permitted a measurement in the range $170 < W_{\gamma^*p} < 260$ GeV. Figure 3.13 shows the dilepton
invariant mass spectra obtained for the elastic sample in the ZEUS analysis.
The e$^+$e$^-$ spectrum shows the asymmetry arising from final-state radiative
effects. A polynomial approximation to this shape was folded with the Gaussian peak and with a flat background arising from Bethe–Heitler processes to
produce the total fit result shown. No such radiative correction was required
for the muon spectrum. The widths of the spectra were consistent with the

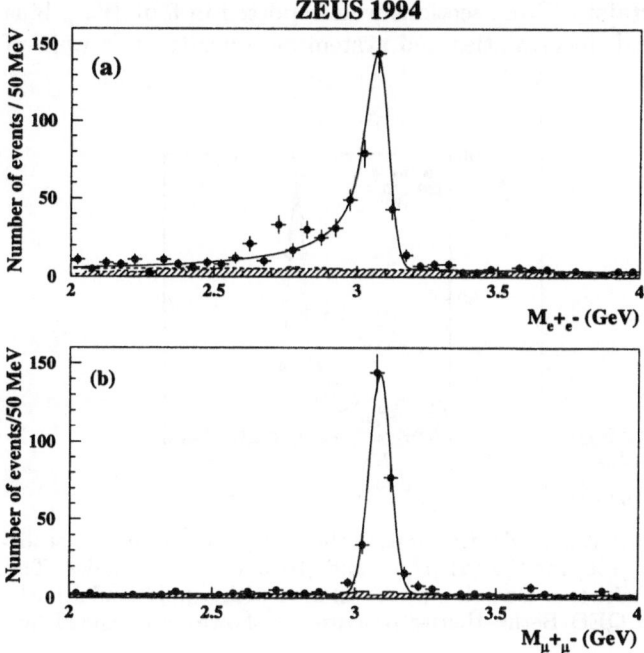

Fig. 3.13. The dilepton invariant mass spectra in the J/ψ mass region in the ZEUS elastic photoproduction study [135] for the (a) e^+e^- and (b) $\mu^+\mu^-$ decay modes. The e^+e^- spectrum includes elastic events only and shows the asymmetry arising from final-state radiative effects. The radiative mass spectrum was convoluted with a Gaussian resolution function and added to a flat background to produce the total fit result shown. A satisfactory fit to the distribution was obtained with a function similar to that employed for the electron final state but omitting the radiative contribution

detector resolution of 30 MeV. The estimate for the background from proton dissociative processes was 30% with an uncertainty of about 10% arising from theoretical uncertainties in the simulation of the dissociative processes. The statistical uncertainty in the cross sections reported in this study was about 15%. The systematic uncertainties were about 10% and arose primarily from the uncertainties in trigger efficiency and background estimation.

Figure 3.14 shows the invariant mass spectra for electrons and muons for elastic and inelastic processes together as presented in the H1 study [136], which distinguished elastic ($z = 1$), elastic with proton dissociation ($z \lesssim 1$), and inelastic channels ($z < 0.8$) via energy depositions in the forward detectors. A mass resolution of ≈ 70 MeV was indicated by the fit results, compatible with expectations from detector simulations. The processes could be distinguished such that background estimates were about 10% with an uncertainty of similar magnitude. This background uncertainty, together with an uncertainty of similar magnitude in the trigger efficiency, dominated the total systematic uncertainty. Cross sections were reported in four W_{γ^*p} bins such that the statistical uncertainties and systematic uncertainties were of comparable magnitude.

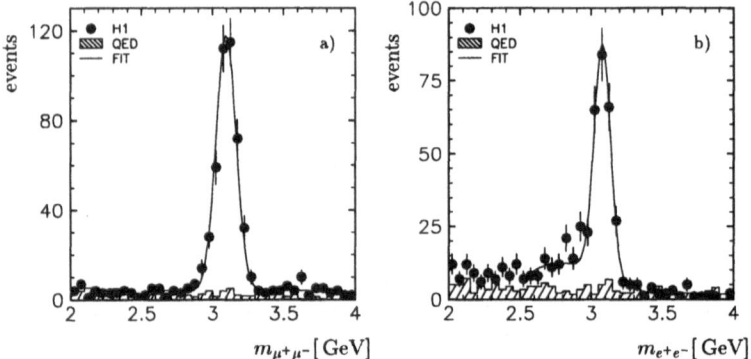

Fig. 3.14. Dilepton invariant mass distributions in the J/ψ photoproduction study by the H1 collaboration [136] for the (a) $\mu^+\mu^-$ and (b) e^+e^- decay modes. The curves are the result of a fit to a Gaussian peak summed with a polynomial background. A simulation of QED Bethe–Heitler processes is shown as a shaded histogram

Clear J/ψ signals at $Q^2 > 7\,\mathrm{GeV}^2$ have been observed in both the H1 [124] and ZEUS [119], experiments, but the conclusions of these studies were severely limited by the available statistics. Results on the Q^2 dependence of the cross section and on the rate relative to that for the ρ^0 meson at high Q^2 will be discussed below.

$\psi(2S) \rightarrow J/\psi\,\pi^+\pi^-$. An observation of photoproduction of the $\psi(2S)$ in its decay to $J/\psi\,\pi^+\pi^-$ has been reported by the H1 collaboration [137]. The requirement of four charged tracks restricted the acceptance to the kinematic region $z > 0.95$. The acceptance was about 7% in the $\mu^+\mu^-$ decay channel and about 5% in the e^+e^- channel. Employing a sample of about two dozen events, they report a result for the ratio of elastic $\psi(2S)$ photoproduction to that for the J/ψ of 0.16 ± 0.06 at a center-of-mass energy of 80 GeV.

3.2.3 Elastic Production Cross Sections

This section summarizes the HERA measurements for the exclusive vector-meson production cross sections. We discuss the dependence of these cross sections on the photon-proton center-of-mass energy, on the photon virtuality, and on the momentum transferred at the proton vertex.

The Energy Dependence and Q^2 Dependence of the Cross Section. The various topical models for the dynamics of exclusive vector-meson production have been outlined in Chap. 2. We turn now to the cross section

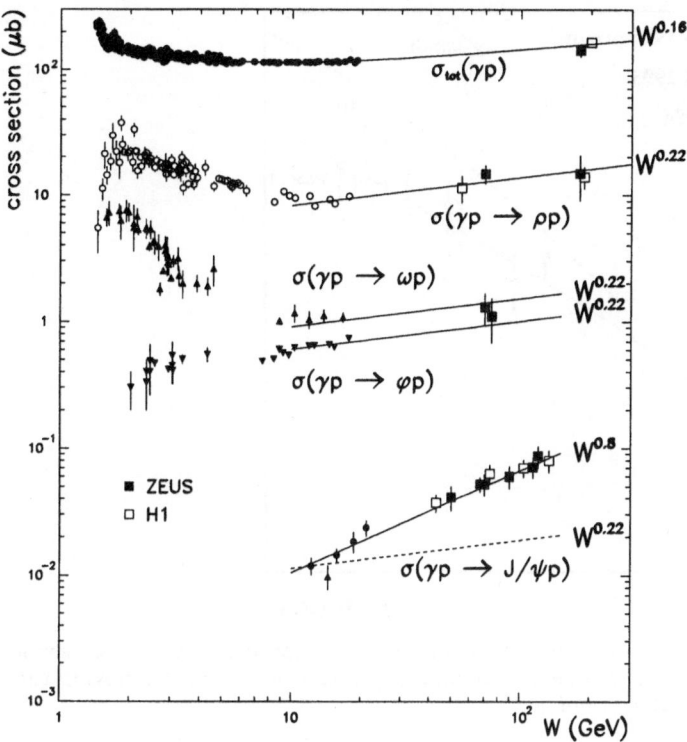

Fig. 3.15. The energy dependence of the ρ^0, ω, ϕ and J/ψ exclusive photoproduction cross sections measured at HERA, compared to measurements at lower energy and to the total photoproduction cross section. The lines illustrate a comparison of various power-law energy dependences at high energy

measurements and discuss them in the context of the theoretical models. Figure 3.15 shows the photoproduction cross sections for elastic ρ^0, ω, ϕ, and J/ψ production measured in the two HERA experiments and compares them to measurements at lower energy and to the total inclusive γ^*p cross section. The stark differences between the production of the ω and the ϕ states observed in the low-energy resonance region (attributable to the fact that only Pomeron exchange contributes to ϕ production [148]) are absent in the energy region measured at HERA. The weak energy dependence associated with soft Pomeron exchange applies to the ρ^0, ω, and ϕ cross sections. The J/ψ cross section, however, shows a distinctly stronger energy dependence, motivating a modified description of the underlying physical process. Modifications to the Pomeron-exchange process have been suggested to account for the stronger energy dependence [80, 81, 82]. Perturbative QCD calculations have also been proposed to describe these measurements [88]. They are of particular interest because they entail a strong sensitivity to the gluon density in the proton (see Chap. 2).

Figure 3.16 compares the new J/ψ measurements to those at lower energy [66] and to the results of pQCD calculations for three gluon density

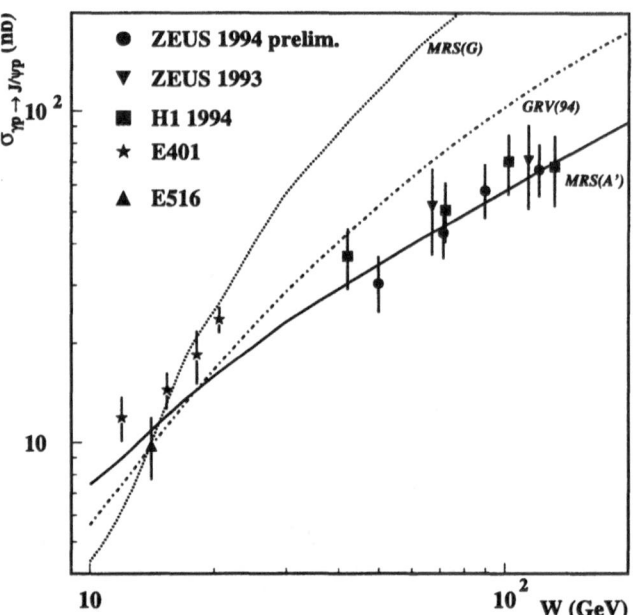

Fig. 3.16. HERA measurements of the elastic J/ψ photoproduction cross section compared to those at lower energy [66] and to pQCD calculations employing three gluon density parametrizations [88]

parametrizations, which exhibit significantly different slopes in the HERA energy region and are only weakly constrained by other measurements.

Figures 3.17 and 3.18 show the H1 and ZEUS results for the energy dependences of the ρ^0 and ϕ production cross sections for Q^2 values between 3.7 and 20 GeV^2, compared to various power-law assumptions. The value of the power used to represent expectations based on the phenomenological Pomeron exchange was obtained using the parameters in (2.2.10) and averaging over the t distribution in the data. Measurements by the NMC collaboration in fixed-target muon scattering [59] are also presented, allowing comparison over an order of magnitude in energy. The Q^2 dependence observed in the NMC results was used to scale the cross sections to the same average Q^2 values as in the ZEUS measurements. In order to account for the different average value of the photon polarization parameter ε in the NMC data ($\langle\varepsilon\rangle \approx 0.75$), the total exclusive cross section is calculated as follows:

$$\sigma^{\mathrm{T}}_{\gamma^*\mathrm{p}\to\mathrm{Vp}} + \sigma^{\mathrm{L}}_{\gamma^*\mathrm{p}\to\mathrm{Vp}} = \left(\frac{1+R}{1+\varepsilon R}\right)\sigma^{\mathrm{tot}}_{\gamma^*\mathrm{p}\to\mathrm{Vp}}, \tag{3.2.37}$$

with

$$R = \frac{\sigma^{\mathrm{L}}_{\gamma^*\mathrm{p}\to\mathrm{Vp}}}{\sigma^{\mathrm{T}}_{\gamma^*\mathrm{p}\to\mathrm{Vp}}}, \tag{3.2.38}$$

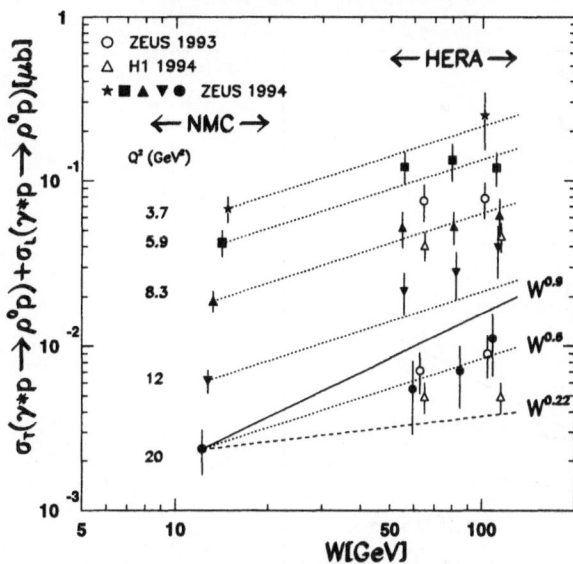

Fig. 3.17. The elastic ρ^0 cross sections from [119]. The dashed line indicates the $W^{0.22}$ dependence expected from the exchange of the phenomenological Pomeron, while the dotted lines indicate the stronger dependence more consistent with the data and the solid line an approximate upper bound

where the value of R is determined via the helicity analyses described in Sect. 3.2.5 under the assumption that s-channel helicity is conserved in the photon/vector meson transition. The photoproduction results included for comparison have been used to obtain the total cross section assuming $R=0$. While the uncertainties in the cross sections prevent a firm conclusion, there does appear to be evidence for a steeper energy dependence at high Q^2 than in photoproduction, even for these lighter-mass vector mesons. In the context of the perturbative QCD calculations such an observation leads to the conclusion that the hard scale may indeed be set either by the mass of the vector meson, as for the J/ψ in photoproduction, or by the photon virtuality.

The most accurate measurements of the Q^2 dependence of the vector-meson production cross sections have been presented for the ρ^0, owing to the greater statistical precision. This dependence is interesting as a test of the perturbative QCD predictions [86, 89]. Calculations of the cross section for longitudinally polarized photons to form vector mesons in interactions where the photon virtuality sets the hard scale result in Q^2 dependences weaker than Q^{-6}, since the Q^{-6} factor (see Chap. 2) is mitigated by the scale-breaking Q^2 dependence in the gluon distribution functions and in the running strong coupling constant. The models based on Regge phenomenology predict the cross section to scale as Q^{-6} [76].

1994 ZEUS

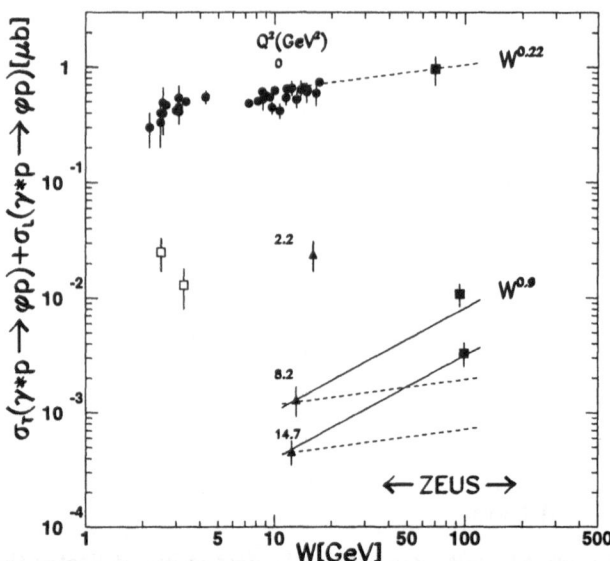

Fig. 3.18. The elastic ϕ cross sections from [119]. The solid circles, open squares and solid triangles represent low-energy results from fixed-target experiments. The results from the ZEUS collaboration are shown as solid squares. The measurements at high Q^2 have additional correlated systematic uncertainties of 32% which are not shown. The NMC measurements *(solid triangles)* have been scaled to the Q^2 values of the ZEUS measurements and have 20% normalization uncertainties which are not shown. The dashed lines indicate an energy dependence consistent with expectations of Regge phenomenology, while the solid lines indicate the power-law dependence $W^{0.9}$ more consistent with the measurements at high Q^2

Figure 3.19 shows results for the Q^2 dependence in the elastic ρ^0 production cross section $\sigma^{tot}_{\gamma^*p \to Vp}$ as measured by the ZEUS collaboration [119] for various values of the center-of-mass energy and parametrized as Q^{-2a}:

$$a = 2.52 \pm 0.28 \text{ (stat)} \pm 0.24 \text{ (sys)} \qquad \langle W \rangle = 56 \text{ GeV},$$

$$a = 2.37 \pm 0.28 \text{ (stat)} \pm 0.25 \text{ (sys)} \qquad \langle W \rangle = 81 \text{ GeV},$$

$$a = 1.84 \pm 0.19 \text{ (stat)} \pm 0.24 \text{ (sys)} \qquad \langle W \rangle = 110 \text{ GeV}.$$

A result for the Q^2 dependence in elastic ϕ production was also presented:

$$a = 2.0 \pm 0.6 \text{ (stat)} \qquad \langle W \rangle = 96 \text{ GeV}.$$

The H1 collaboration has published results for the Q^2 dependence for ρ^0 mesons [124]:

$$a = 2.5 \pm 0.5 \text{ (stat)} \pm 0.2 \text{ (sys)} \qquad \langle W \rangle = 81 \text{ GeV},$$

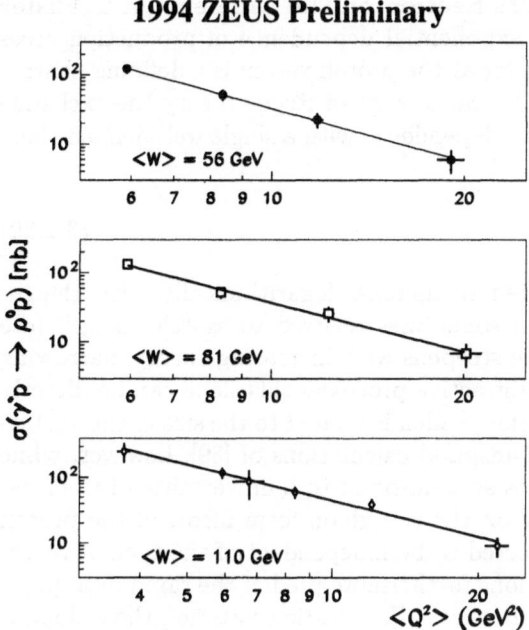

Fig. 3.19. The Q^2 dependence of the elastic ρ^0 cross section in three ranges of $W_{\gamma^* p}$ from the preliminary ZEUS results using data from 1994 [119]. The solid circles, open squares, and open diamonds represent data restricted to the kinematic regions $0.02 < y < 0.05$, $0.05 < y < 0.10$, and $0.10 < y < 0.20$ respectively. The points at Q^2 values of 3.7 and 6.9 GeV^2 in the lower plot were obtained from a data sample where the interaction vertex was shifted in the forward direction to gain acceptance at lower values of Q^2

and ϕ mesons [133] as well:

$$a = 2.0 \pm 0.6 \text{ (stat)} \pm 0.2 \text{ (sys)} \qquad \langle W \rangle = 88 \text{ GeV}.$$

The H1 collaboration further published a result for the J/ψ which combines their measurements at high Q^2 with a preliminary ZEUS photoproduction result and uses the parametrization $(Q^2 + M^2_{J/\psi})^{-a}$ to obtain [124]:

$$a = 1.9 \pm 0.3 \text{ (stat)} \qquad \langle W \rangle = 92 \text{ GeV}.$$

These results do not allow an unambiguous distinction between the Q^2 dependences for the various vector mesons. However, there is evidence that the power-law dependence is weaker than Q^{-6}, consistent with pQCD calculations for longitudinal photons, as discussed by Frankfurt et al. [91], for example. There are also indications that the Q^2 dependence weakens as the energy increases. A straightforward comparison with the calculations is not yet possible, since the measurements consist of a mixture of contributions from transversely and longitudinally polarized photons (see Sect. 3.2.5).

The Dependence of the Cross Section on the Momentum Transfer at the Proton Vertex. The exponential dependence of production cross sections on the momentum transfer at the proton vertex is a defining characteristic of diffractive processes. In the context of Regge theory the exchange of a single trajectory results in a t dependence with a single well-defined slope parameter b:

$$\frac{d\sigma}{d|t|} = A\, e^{-b|t|}. \tag{3.2.39}$$

This slope parameter is expected to increase logarithmically with the interaction energy, a phenomenon sometimes referred to as "shrinkage" (see Chap. 2), since the t dependence steepens with increasing energy, narrowing the foward diffractive peak. In diffractive processes this slope can be directly related to the size of the interaction, which is related to the size of the vector-meson wave packet. The pQCD-inspired calculations of [89], however, while predicting slight shrinkage effects as well, point to a universality of the t dependence, since it depends only on the two-gluon form factor of the proton at high Q^2 and is therefore expected to be independent of the type of vector meson produced. These calculations furthermore predict the rapid disappearance of shrinkage effects as Q^2 increases. For elastic scattering the t slope is limited to values greater than the proton size, that is, to values greater than about $4\,\mathrm{GeV}^{-2}$.

The ZEUS Leading Proton Spectrometer provided a direct measurement of the proton transverse momentum for the data recorded in 1994, achieving a resolution of $0.01\,\mathrm{GeV}^2$ in the squared momentum transfer [19] and reducing the background from processes with proton dissociation to less than 0.5%. The result for the slope parameter in elastic ρ^0 photoproduction was

$$b = 9.8 \pm 0.8\ (\mathrm{stat}) \pm 1.1\ (\mathrm{sys})\ \mathrm{GeV}^{-2},$$

where the systematic error was primarily determined by the acceptance of the LPS detector and the influence of the intrinsic transverse momentum of the proton beam.

All other H1 and ZEUS analyses use the total transverse momentum observed in the final state as an estimator for the momentum transferred to the proton. For analyses in which the scattered electron was not detected, the Q^2 dependence in the relationship between p_T^2 and t was accounted for in the final results via Monte Carlo simulation and unfolding techniques. The most accurate measurements of the slope parameter from HERA have been presented by the ZEUS collaboration using a high-statistics $\pi^+\pi^-$ sample from the 1994 data-taking period [126] where neither the electron nor the proton was detected in the final state. A deviation from a purely exponential dependence was observed, motivating the parametrization

$$\frac{d\sigma}{d|t|} = A\, e^{-b_{\pi\pi}|t| + c_{\pi\pi}t^2}. \tag{3.2.40}$$

Figure 3.20 shows the t dependence of the measured cross section together with the results of the fit:

$$b_{\pi\pi} = 11.6 \pm 0.2 \text{ (stat) GeV}^{-2}, \qquad c_{\pi\pi} = 4.2 \pm 0.5 \text{ (stat) GeV}^{-4}.$$

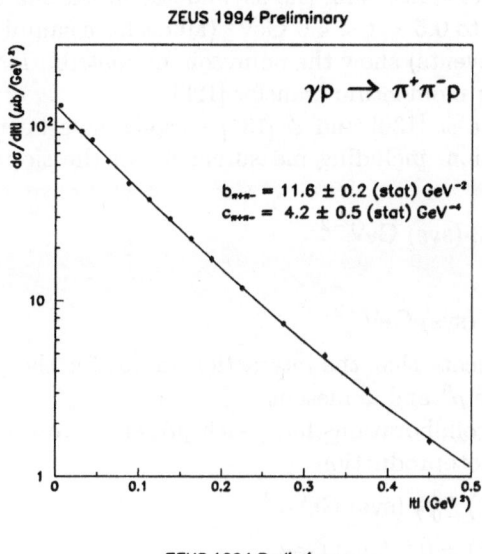

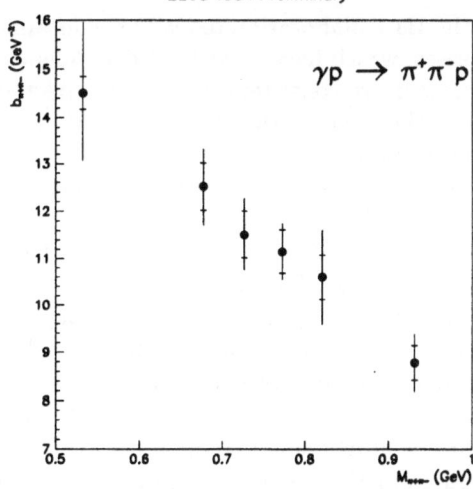

Fig. 3.20. The t dependence of the elastic $\pi^+\pi^-$ photoproduction cross section from [126]. Results for the fit parameters described in the text are shown in the upper plot. The lower plot shows the linear slope parameter value as a function of the dipion invariant mass. The inner error bars represent the statistical uncertainties and the outer error bars represent the quadratic sum of statistical and systematic uncertainties

The slope parameter clearly depends on the range of dipion invariant mass $(M_{\pi\pi})$ included in the fit, an effect observed in earlier photoproduction studies [50, 149]. As described in Sect. 3.2.2 the mass spectrum is influenced by the contribution of an interference term from a nonresonant dipion background, leading to the dependence of the slope parameter on $M_{\pi\pi}$ shown in Fig. 3.20, and to the necessity for the nonlinear term in the exponent in (3.2.40). Further preliminary results concerning the asymmetry in the mass spectrum which extend the t range to $0.5 < t < 4.0$ GeV^2 (albeit for a sample dominated by proton-dissociative events) show the nonresonant contribution to decrease sharply with increasing momentum transfer [128].

Photoproduction results for the ω [130] and ϕ [131] mesons have been published by the ZEUS collaboration, including measurements of the slope parameter b:

$$b = 10.0 \pm 1.2 \text{ (stat)} \pm 1.3 \text{ (sys) GeV}^{-2}$$

for the ω meson, and

$$b = 7.3 \pm 1.0 \text{ (stat)} \pm 0.8 \text{ (sys) GeV}^{-2}$$

for the ϕ meson. These results indicate that the interaction radius for the ϕ meson is smaller than those for the ρ^0 and ω mesons.

The ZEUS [135] and H1 [136] collaborations have each presented results on the slope parameter for J/ψ photoproduction:

$$\text{ZEUS}: b = 4.0 \pm 0.4 \text{ (stat)} \, ^{+0.6}_{-0.7} \text{ (sys) GeV}^{-2},$$

$$\text{H1}: b = 4.0 \pm 0.2 \text{ (stat)} \pm 0.2 \text{ (sys) GeV}^{-2}.$$

The forward detectors were used by the H1 collaboration to exclude contamination from proton dissociative processes, which have a weaker t dependence. These small values for the slope parameter are consistent with that expected from the proton alone, indicating that the contribution from the J/ψ wave function is smaller than that from the proton.

All HERA studies of the photoproduction slope parameters covered in this book lack the precision necessary to identify deviations from the logarithmic energy dependence expected in soft diffractive processes governed by exchange of the phenomenological Pomeron. The comparisons to measurements at lower energy also lack such precision, despite the wide range of energy covered, owing to contributions from the uncertainties in the earlier measurements. Probably the best hope for future information on the topic of shrinkage is measurement at HERA of the slope parameter over an extended range of energy (50–200 GeV) with an accuracy small compared to the rise of approximately 0.7 GeV^{-2} expected from the phenomenological Pomeron trajectory over this range.

The status of the slope parameter measurements at high Q^2 for ρ^0 production [119] is shown in Fig. 3.21. The precision is limited by statistics, and is hence significantly better for the ρ^0 meson than for the ϕ and J/ψ mesons, for which slope determinations have also been presented. The preliminary ZEUS results from the 1994 data-taking period cover the kinematic

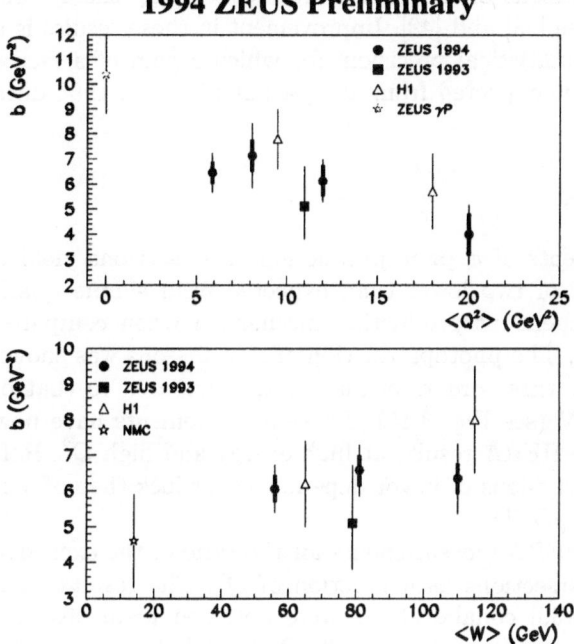

Fig. 3.21. A comparison of slope parameter determinations at high Q^2 [119]. Preliminary results from the ZEUS collaboration from the 1994 data-taking period are shown, covering the energy range $42 < W_{\gamma^* p} < 134$ GeV in the upper plot and the range $5 < Q^2 < 30$ GeV^2 in the lower plot. The measurements are compared to published results from the H1 [124] and NMC [59] experiments, and to the ZEUS photoproduction result from [85]. The inner error bars on the ZEUS points show the contribution from statistics, and the outer error bars represent the quadratic sum of statistical and systematic uncertainties, excluding a correlated contribution of 1.5 GeV^{-2}

region $5 < Q^2 < 30$ GeV^2 and $42 < W_{\gamma^* p} < 134$ GeV. The slopes were determined by fits to the distributions in $(\boldsymbol{p}_T(e) + \boldsymbol{p}_T(\rho^0))^2$. The inner error bars show the contribution from statistics, and the outer error bars represent the quadratic sum of statistical and systematic uncertainties, excluding a correlated contribution of 1.5 GeV^{-2}. The measurements of the H1 [124] and NMC [59] experiments are also shown for comparison, as is the ZEUS photoproduction result. The preliminary conclusions from the observed Q^2 and $W_{\gamma^* p}$ dependences are that there is evidence for a decrease in the slope parameter with increasing Q^2, but that the present level of uncertainty excludes confirmation of any shrinkage effect in the energy dependence. The decrease of the slope with Q^2 is indicative of a reduction in size of the $q\bar{q}$ state with Q^2. The value for the slope of about 4 GeV^{-2} at high Q^2 is comparable to the size of the proton alone as measured in its hadronic interactions, leading to the conclusion that the $q\bar{q}$ state derived from the photon is smaller than the proton. For discussions of theoretical ideas concerning the Q^2 dependence of photon size, see [23] and [49]. Improvement in these results is a matter of increasing their statistical precision, for which a gain of a factor of approximately five can be expected from analyses of 1995 and 1996 data presently in progress.

3.2.4 Production Ratios

The low-energy measurements of ϕ photoproduction cross sections yielded results more than a factor of two lower than expected from simple quark counting and a flavor-independent production mechanism when compared to ρ^0 photoproduction [49]. The photoproduction of J/ψ mesons was shown to be suppressed by nearly three orders of magnitude compared to that of the ρ^0 at $\langle W_{\gamma^*p} \rangle = 15$ GeV (see Fig. 3.15). These measurements have now been complemented by the HERA results at high energy and high Q^2. References [72, 89] include discussions of flavor dependence, or lack thereof, expected in models based on pQCD.

Figure 3.22 shows the HERA measurements for the ratio of the exclusive ϕ and ρ^0 production cross sections as a function of Q^2. The results from the ZEUS [132] and H1 [133] collaborations were obtained at an average photon–proton center-of-mass energy of about 90 GeV, while those from the fixed-target muon experiments EMC [57] and NMC [59] were obtained at

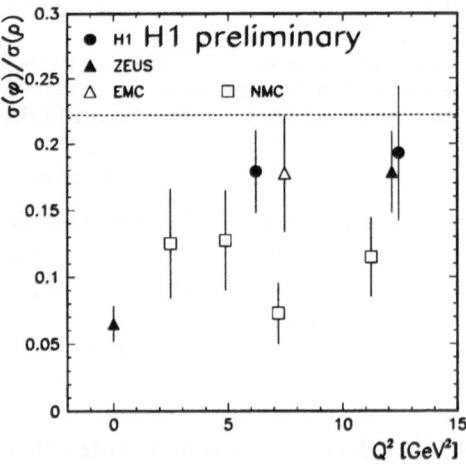

Fig. 3.22. The Q^2 dependence of the ϕ-to-ρ^0 production ratio. The results from the ZEUS [132] and H1 [133] collaborations were obtained at an average photon–proton center-of-mass energy of about 90 GeV, while those from the fixed-target muon experiments EMC and NMC (shown as the open squares and triangle) were obtained at 15 GeV. The dashed line corresponds to the value for the ratio of 2/9 derived from a simple quark-counting rule under the assumption of a flavor-independent production process

15 GeV. The data support a conclusion that at high energy and high Q^2 the value of 2/9 expected from a quark-counting rule and flavor symmetry is reached. The ZEUS collaboration has included a discussion of ϕ and ω exclusive photoproduction relative to that for the ρ^0 in their article on ω photoproduction [130], concluding that the observed energy dependence is consistent with expectations arising from a diffractive process mediated by Pomeron exchange.

As discussed in Sect. 3.2.3, the energy dependence observed in the elastic J/ψ photoproduction cross section is much stronger than that observed in ρ^0 photoproduction. The J/ψ cross section rises by nearly an order of magnitude as $W_{\gamma^* p}$ increases from 10 to 100 GeV, while the energy dependence of the ρ^0 cross section is consistent with the weak energy dependence typical of soft Pomeron exchange. The $J/\psi/\rho^0$ photoproduction ratio is thus steeply rising with energy, though production of the J/ψ is still relatively suppressed by two orders of magnitude at HERA energies. This suppression stands in sharp contrast to the results presented at high Q^2. The ZEUS collaboration has presented a result for the ratio of $(0.35 \pm 0.11 \text{ (stat)})$ for the range $7 < Q^2 < 25$ GeV^2 [119], confirming the earlier measurement of comparable cross sections for exclusive ρ^0 and J/ψ production at high Q^2 published by the H1 collaboration [124]. These results represent further evidence for an underlying flavor-independent production process at high Q^2. In the context of QCD, the dominance of ρ^0 production at low Q^2 can be interpreted as being due to nonperturbative effects associated with the large value of the strong coupling constant in the absence of a hard scale.

3.2.5 Helicity Analyses

The ratio of the vector-meson production cross section for longitudinally polarized photons to that for transversely polarized photons is a quantity for which there are specific predictions which characterize several of the models employed to describe the high-Q^2 measurements. A firm prediction of the pQCD calculations is that the process is dominated by the production of longitudinally polarized vector mesons from longitudinally polarized photons at high Q^2 [89, 75]. The applicability of the calculations depends crucially on this prediction, since the production by transversely polarized photons is not included. Models based on Pomeron exchange similarly predict the predominance of longitudinally polarized vector-meson production at high Q^2 [76, 150]. The color dipole model of [95] also offers a specific prediction for the Q^2 dependence of the cross section ratio: the ratio of the longitudinal contribution to the transverse contribution is expected to rise more slowly than Q^2/M_V^2.

The polarization of the vector meson is experimentally accessible via the decay angular distributions, accurately measured in the central tracking chambers of the HERA experiments. A general geometrical description of the process [74] (see Fig. 3.23) involves the definition of three planes in the rest

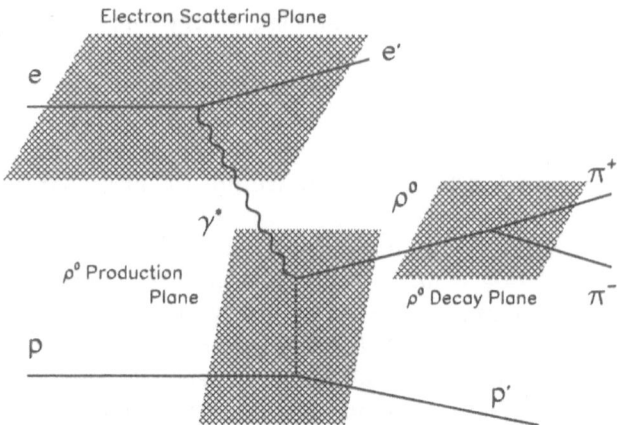

Fig. 3.23. Definitions of the planes used in the helicity analyses

frame of the ρ^0 meson: (1) the electron scattering plane, (2) the plane containing the momentum vectors of the initial- and final-state protons and that of the photon (called the production plane), and (3) the vector-meson decay plane. The vector-meson production mechanism then defines which choice of reference frame is best adapted to a simple description of the process. Bubble chamber studies of ρ^0 photoproduction [50, 149] have shown that the spin-density matrix elements defined in the Gottfried–Jackson and Adair reference systems were strongly dependent on t, thus excluding t-channel helicity-conserving and spin-independent processes as candidates for the dominant underlying production mechanism. (Reference [50] includes a nice discussion of these reference frames.) The s-channel helicity frame (the ρ^0 meson rest frame with the quantization axis chosen as the direction of the ρ^0 meson in the γp center-of-mass system) proved to yield spin-density matrix elements independent of the momentum transfer. This result pointed to an underlying process in which helicity is conserved in the photon–vector-meson transition for amplitudes defined in the s-channel (SCHC). (Theoretical discussions of s-channel helicity conservation can be found in [151, 152].) A streamer chamber experiment studying ρ^0 electroproduction on a hydrogen target at DESY also concluded the validity of SCHC [51]. Such results have led to the helicity frame being chosen as reference in the HERA studies at both low and high Q^2, though present experimental precision has not allowed a full spin-density analysis of the type performed for the low-energy measurements. The spin-density matrix can be decomposed into even and odd (referred to as "natural" and "unnatural") parity-exchange components when the amplitudes are expressed in the t-channel [74, 153]. Interference between these two components decreases with energy. Linear photon polarization allows the experimental distinction of natural and unnatural parity-exchange components. Together with the assumption of SCHC, the transformation properties of the underlying production process under t-channel parity exchange completely determine the spin-density matrix elements.

A complete characterization of the process (see, for example, [153] and [154]) requires the definition of three angles in the vector-meson rest frame. Two of these are canonically chosen as the polar (θ_h) and azimuthal (ϕ_h) angles of the momentum of the positively charged vector-meson decay product in the coordinate system where the vector-meson momentum in the γp center-of-mass frame is the z axis and the y axis is perpendicular to the vector-meson production plane. The third is the angle between the vector-meson production plane and the electron scattering plane (Φ_h). For small values of the momentum transfer at the proton vertex, $|t| \gtrsim |t|_{min}$, the photon and vector meson are nearly collinear and the angle ϕ_h becomes ill-defined and poorly measured. The assumption of s-channel helicity conservation constrains the azimuthal distribution to depend solely on the difference of the two azimuthal angles: $\psi_h = \Phi_h - \phi_h$. The angle Φ_h can be determined only if the direction of the final-state electron momentum is measured, so the untagged photoproduction experiments necessarily average over it. Furthermore, the angle ϕ_h can be determined only if the photon direction of flight is known, but in practice it turns out that for the photoproduction processes studied at HERA the flight direction of the photon is a good approximation to that of the initial-state electron, so the azimuthal decay angular distribution can be determined even when the final-state electron is not detected.

The three-dimensional angular distribution for the case of the unpolarized (or transversely polarized) electron beam at HERA is given by [153]:

$$W(\cos\theta_h, \phi_h, \Phi_h) =$$

$$
\begin{aligned}
\frac{3}{4\pi}\Bigg[& \left(\frac{1}{2}\left(1 - r_{00}^{04}\right) + \frac{1}{2}\left(3r_{00}^{04} - 1\right)\cos^2\theta_h \right. \\
& \left. - \sqrt{2}\,\mathrm{Re}\,r_{10}^{04}\,\sin 2\theta_h \cos\phi_h - r_{1-1}^{04}\sin^2\theta_h \cos 2\phi_h \right) \\
& - \varepsilon\cos 2\Phi_h(\; r_{11}^1 \sin^2\theta_h + r_{00}^1\cos^2\theta_h \\
& \qquad\qquad - \sqrt{2}\,\mathrm{Re}\,r_{10}^1\,\sin 2\theta_h \cos\phi_h \\
& \qquad\qquad - r_{1-1}^1 \sin^2\theta_h \cos 2\phi_h\;) \\
& - \varepsilon\sin 2\Phi_h(\; \sqrt{2}\,\mathrm{Im}\,r_{10}^2\,\sin 2\theta_h \sin\phi_h \\
& \qquad\qquad + \mathrm{Im}\,r_{1-1}^2 \sin^2\theta_h \sin 2\phi_h\;) \\
& + \sqrt{2\varepsilon(1+\varepsilon)}\cos\Phi_h(\; r_{11}^5 \sin^2\theta_h + r_{00}^5 \cos^2\theta_h \\
& \qquad\qquad - \sqrt{2}\,\mathrm{Re}\,r_{10}^5\,\sin 2\theta_h \cos\phi_h \\
& \qquad\qquad - r_{1-1}^5 \sin^2\theta_h \cos 2\phi_h\;) \\
& \qquad\qquad \sqrt{2}\,\mathrm{Im}\,r_{10}^6\,\sin 2\theta_h \sin\phi_h \\
& + \sqrt{2\varepsilon(1+\varepsilon)}\sin\Phi_h(\; + \mathrm{Im}\,r_{1-1}^6 \sin^2\theta_h \sin 2\phi_h)\Bigg],
\end{aligned}
\tag{3.2.41}
$$

where ε is the virtual photon polarization parameter defined in (3.1.18). The subscripts i and k run over the three possible helicity states -1, 0, and $+1$. The quantities $r_{ik}^{\alpha}, r_{ik}^{04}$, are linear combinations of the spin-density matrix elements ϱ_{ik}^{α}, ($\alpha=0$, 1, 2, 4, 5, 6), which themselves depend on the helicity amplitudes and are sensitive to the dynamics of the production process:

$$r_{ik}^{04} = \frac{\varrho_{ik}^0 + \varepsilon R \varrho_{ik}^4}{1 + \varepsilon R} \tag{3.2.42}$$

$$r_{ik}^{\alpha} = \begin{cases} \dfrac{\varrho_{ik}^{\alpha}}{1 + \varepsilon R}, & \alpha = 1, 2 \\[2ex] \dfrac{\sqrt{R}\,\varrho_{ik}^{\alpha}}{1 + \varepsilon R}, & \alpha = 5, 6 \end{cases} \tag{3.2.43}$$

The superscript 0 corresponds to the case of unpolarized transverse photons; the superscripts 1 and 2 indicate terms arising in the case of linearly polarized transverse photons; the superscript 4 represents the contribution from longitudinally polarized photons; and the terms with superscripts 5 and 6 indicate the interference between the longitudinal and transverse amplitudes. Terms arising from circularly polarized photons (longitudinally polarized initial-state leptons) carry the superscript 3 and are accessible in fixed-target muoproduction, since the muons are produced with longitudinal polarization via pion decay. The variable R is the ratio of the elastic ρ^0 production cross section for longitudinal photons to that for transverse photons.

If the helicity amplitudes obey SCHC and natural parity exchange, then the only nonzero density matrix elements are $\varrho_{00}^4 = 1, \operatorname{Re} \varrho_{1-1}^1 = -\operatorname{Im} \varrho_{1-1}^2 = \frac{1}{2}, \operatorname{Re} \varrho_{10}^5 = -\operatorname{Im} \varrho_{10}^6 = -\cos\delta/\sqrt{8}$ and the angular distribution reduces to

$$W(\cos\theta_{\mathrm{h}}, \psi_{\mathrm{h}}) = \frac{3}{8\pi} \frac{1}{1 + \varepsilon R} \left[\sin^2\theta_{\mathrm{h}}(1 + \varepsilon \cos 2\psi_{\mathrm{h}}) + 2\varepsilon R \cos^2\theta_{\mathrm{h}} \right.$$
$$\left. - \sqrt{2R\varepsilon(1 + \varepsilon)} \cos\delta \sin 2\theta_{\mathrm{h}} \cos\psi_{\mathrm{h}} \right], \tag{3.2.44}$$

where the parameter δ is the relative phase angle between the longitudinal and transverse amplitudes. The value of R can be deduced directly from the value of r_{00}^{04} determined by the polar angle dependence when averaging over the azimuthal dependence:

$$R = \frac{1}{\varepsilon} \frac{r_{00}^{04}}{1 - r_{00}^{04}}. \tag{3.2.45}$$

The first measurements at HERA of the spin-density matrix elements derived from vector-meson decay angular distributions [85] were based on the observation of ρ^0 mesons produced in photoproduction by the ZEUS collaboration. The polar and azimuthal angular distributions are shown in Fig. 3.24. The observed $\sin^2 \theta_h$ dependence indicates that the ρ^0 mesons were produced in a transversely polarized state, consistent with the expectations arising from SCHC. The matrix element was determined to be $r_{00}^{04} = (0.055 \pm 0.028)$. The final-state electron is not measured and the resulting average over Φ_h eliminates the possibility of a contribution from r_{1-1}^1, so the hypothesis of a contribution from r_{1-1}^{04} to the azimuthal distribution was tested. It was found to be small: $r_{1-1}^{04} = (0.008 \pm 0.014)$, consistent with the value derived from SCHC, since SCHC requires the matrix elements ϱ_{1-1}^0 and ϱ_{1-1}^4 to vanish. From the value of r_{00}^{04} a value for R of (0.06 ± 0.03) was calculated according to (3.2.45), in agreement with the expectation in the vector-dominance model of a value of 0.1.

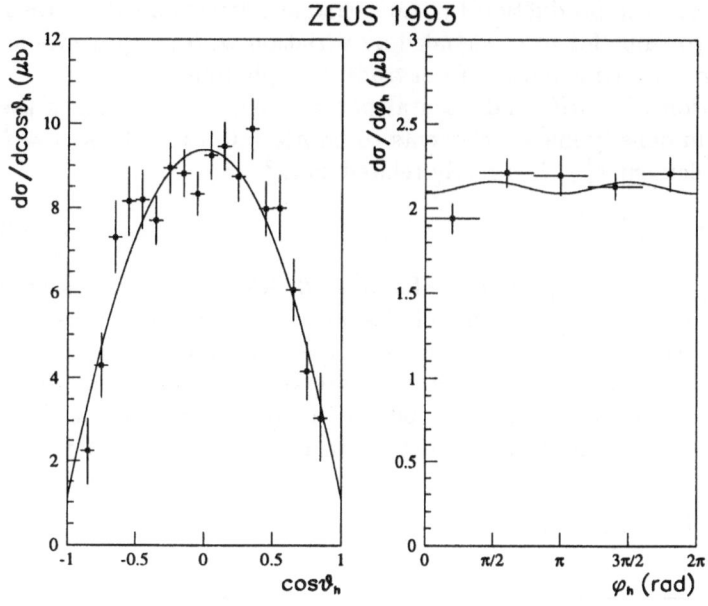

Fig. 3.24. The helicity angle distributions observed in photoproduction of the ρ^0 meson by the ZEUS collaboration. The dominance of the $\sin^2 \theta_h$ dependence and the lack of ϕ_h dependence are consistent with the expectations based on s-channel helicity conservation

We turn now to the helicity analyses in which the final-state electron was detected, allowing the measurement of the angle between the electron scattering plane and the ρ^0 production plane. Under the assumption of SCHC the azimuthal distribution is dependent only upon the difference of ϕ_h and Φ_h. Calculating the projections from (3.2.44) we write the angular distributions

$$\frac{1}{N} \frac{dN}{d(\cos \theta_h)} = \frac{3}{4} \left[1 - r_{00}^{04} + (3r_{00}^{04} - 1) \cos^2 \theta_h \right], \tag{3.2.46}$$

$$\frac{1}{N}\frac{\mathrm{d}N}{\mathrm{d}\psi_\mathrm{h}} \;=\; \frac{1}{2\pi}\left(1 + 2\varepsilon\, r^1_{1-1}\cos 2\psi_\mathrm{h}\right).\tag{3.2.47}$$

The spin-density matrix element r^{04}_{00} represents the probability that the vector meson is produced with helicity zero (longitudinally polarized) by either transversely or longitudinally polarized virtual photons. According to (3.2.42) and (3.2.43)

$$r^{04}_{00} = \frac{1}{1+\varepsilon R}\left(\varrho^0_{00} + \varepsilon R\varrho^4_{00}\right),\qquad r^1_{1-1} = \frac{1}{1+\varepsilon R}\,\varrho^1_{1-1},\tag{3.2.48}$$

where ϱ^0_{00} is the probability to produce a longitudinally polarized vector meson from a transverse photon, ϱ^4_{00} is the probability to produce such a longitudinally polarized vector meson from a longitudinal photon, and ϱ^1_{1-1} is sensitive to the interference between nonflip and double-flip amplitudes from transverse photons. Thus one can see that a nonzero value for r^{04}_{00} indicates the production of longitudinally polarized vector mesons, but does not distinguish between the production by transverse and longitudinal photons, whereas a nonzero value for r^1_{1-1}, signaled by variation with ψ_h, provides an unambiguous sign of a contribution from transverse photons.

The assumption of SCHC and natural parity exchange determines the vector-meson spin-density matrix elements to be $\varrho^0_{00} = 0$, $\varrho^4_{00} = 1$, $\varrho^1_{1-1} = \frac{1}{2}$ and the matrix element r^1_{1-1} is directly related to r^{04}_{00}:

$$r^1_{1-1} = \frac{1}{2}(1 - r^{04}_{00}).\tag{3.2.49}$$

A helicity analysis for ρ^0 production for $Q^2 > 8\ \mathrm{GeV}^2$ has been presented by the H1 collaboration [146], which includes a clear indication of a nonzero value for the matrix elements governing both the polar and the azimuthal angular dependences. Figure 3.25 shows the observed angular distributions. The polar angle dependence clearly exhibits the predominantly longitudinal ρ^0 polarization, and a value for R was determined from the value for r^{04}_{00}:

$$r^{04}_{00} \;=\; 0.73 \pm 0.05\ (\text{stat}) \pm 0.02\ (\text{sys}),\qquad R = 2.7^{+0.7}_{-0.5}\ (\text{stat})^{+0.3}_{-0.2}\ (\text{sys}).$$

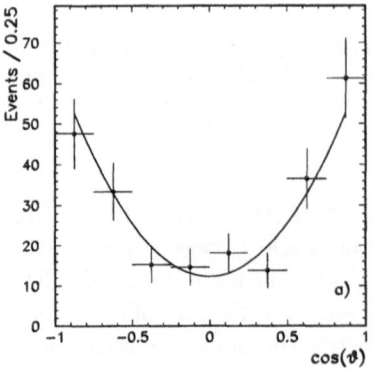 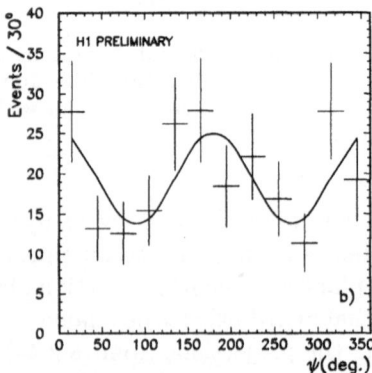

Fig. 3.25. The helicity angle distributions for elastic ρ^0 production obtained by the H1 collaboration for $Q^2 > 8\ \mathrm{GeV}^2$

Since the assumption of SCHC requires r^{04}_{1-1} to vanish, eliminating its contribution to the azimuthal angular distribution (see (3.2.41)), a value for r^1_{1-1} could be determined by a fit to the distribution, with the result

$$r^1_{1-1} \;=\; 0.14 \pm 0.05 \;(\text{stat}) \pm 0.01 \;(\text{sys}).$$

The consistency of this result with the independent expectation from (3.2.49),

$$r^1_{1-1} \;=\; 0.14 \pm 0.03 \;(\text{stat}) \pm 0.01 \;(\text{sys}),$$

supports the validity of the assumption of s-channel helicity conservation with natural parity exchange in the t channel.

The ZEUS collaboration has used data from its beam-pipe calorimeter ($0.25 < Q^2 < 0.85 \;\text{GeV}^2$) to obtain results on ρ^0 production, observing a large longitudinal component, as well as a nonzero value for r^1_{1-1}. This analysis considered two distinct ranges of Q^2, since there is evidence that this region is sensitive to the transition between soft and hard processes (cf. Sect. 3.1), and found a significant increase of the longitudinal contribution with Q^2. Figure 3.26 shows the angular distributions from which the following values for the matrix elements were derived:

$$\langle Q^2 \rangle \;=\; 0.34\,\text{GeV}^2 : \quad r^{04}_{00} = 0.21 \pm 0.03, \quad r^1_{1-1} = 0.38 \pm 0.03,$$

$$R = 0.26 \pm 0.04,$$

$$\langle Q^2 \rangle \;=\; 0.61\,\text{GeV}^2 : \quad r^{04}_{00} = 0.36 \pm 0.04, \quad r^1_{1-1} = 0.31 \pm 0.03,$$

$$R = 0.56 \pm 0.09.$$

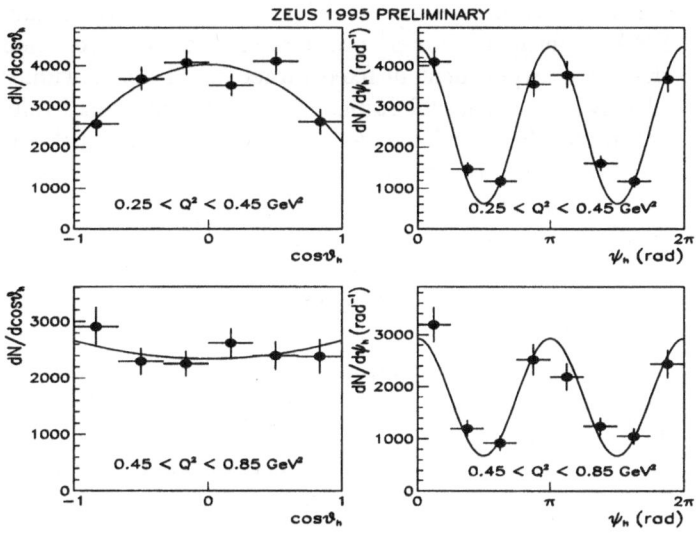

Fig. 3.26. The ρ^0 helicity angle distributions obtained from data recorded in 1995 with the ZEUS beam-pipe calorimeter, separated into two Q^2 regions with average Q^2 values of 0.34 and 0.61 GeV^2

The present status of the helicity angle analyses for the higher-mass vector mesons suffers from more stringent statistical constraints. The ZEUS collaboration has measured matrix elements for ϕ production both in photoproduction [131]:

$$r_{00}^{04} = -0.01 \pm 0.04,$$
$$r_{1-1}^{04} = +0.03 \pm 0.05,$$

and for $\langle Q^2 \rangle = 12.3 \text{ GeV}^2$ [132]:

$$r_{00}^{04} = 0.76_{-0.16}^{+0.11} \text{ (stat)} \pm 0.12 \text{ (sys)}.$$

The photoproduction results are consistent with an underlying diffractive process and the results at both the low and the high Q^2 are consistent with the expectations based on SCHC. Similar conclusions were drawn from the ω photoproduction study [130], where the following values for the matrix elements were determined:

$$r_{00}^{04} = +0.11 \pm 0.08 \text{ (stat)} \pm 0.26 \text{ (sys)},$$
$$r_{1-1}^{04} = -0.04 \pm 0.08 \text{ (stat)} \pm 0.12 \text{ (sys)}.$$

The H1 collaboration has published a helicity analysis for the polar angle in J/ψ photoproduction [136] and find the dependence

$$W(\cos\theta_{\mathrm{h}}) \propto 1 + \cos^2\theta_{\mathrm{h}}, \tag{3.2.50}$$

consistent with the expectations of SCHC for the decay to fermions [152].

We conclude this section with a compilation of results for $R = \sigma^{\mathrm{L}}/\sigma^{\mathrm{T}}$ in the studies at high Q^2 where the predominance of the longitudinal part is an essential prediction of the perturbative models. Figure 3.27 shows the results obtained by the HERA experiments for ρ^0 production and compares them to the result from the NMC experiment in fixed-target muon beam studies. While these results clearly show values for R inconsistent with the small value measured in photoproduction both at low energies and at HERA, conclusive results on the Q^2 dependence of R await studies of higher statistical precision.

The near future is very bright for the continuation of these helicity analyses in elastic vector-meson production at HERA. Soon more results from the high-statistics ρ^0 photoproduction sample from the 1994 and 1995 data-taking periods will become available, which should provide information on the t dependence of the matrix elements, a definitive indicator for the applicability of SCHC. For the high-Q^2 ρ^0 investigations, samples with an order of magnitude greater statistics are now under study. By the end of 1997 the studies in the transition region of $0.2 < Q^2 < 0.9 \text{ GeV}^2$ will also have nearly an order-of-magnitude increase in the number of ρ^0 decays available. By that time precise measurements for the J/ψ should also be possible and the value of R so crucial to the theoretical understanding of these processes will be accessible with better precision over a greater kinematic region and for a wider variety of vector mesons.

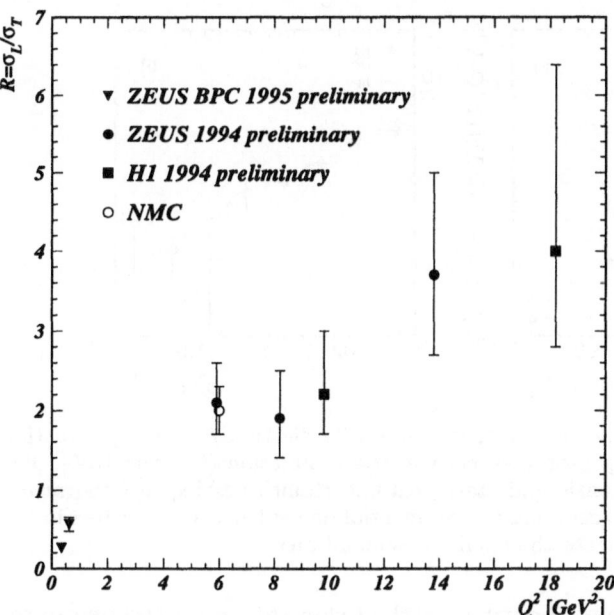

Fig. 3.27. The Q^2 dependence of the ratio of longitudinal to transverse elastic ρ^0 production cross sections from the HERA experiments compared to the fixed-target muon beam result from the NMC experiment [59]. The error bars represent the statistical errors alone

3.3 Vector-Meson Production with Proton Dissociation

The H1 collaboration has isolated a J/ψ data set with proton dissociation [136] in photoproduction as described in Sect. 3.2. The authors report a magnitude for the dissociative cross section comparable to that for the elastic cross section, and a significantly steeper energy dependence in the dissociative sample. The dissociative sample exhibited a much flatter t dependence, as shown in Fig. 3.28. The result of the exponential fit to the p_T^2 slope for the elastic sample was

$$b \;=\; 4.0 \pm 0.2 \,(\text{stat}) \pm 0.2 \,(\text{sys})\; \text{GeV}^{-2},$$

while for the dissociative sample a slope of

$$b \;=\; 1.6 \pm 0.3 \,(\text{stat}) \pm 0.1 \,(\text{sys})\; \text{GeV}^{-2}$$

was found. An additional uncertainty due to the approximation of t with p_T^2 was estimated to be 7%.

The H1 collaboration has also presented results from a study of ρ^0 production with proton dissociation in the Q^2 range $7 < Q^2 < 36\ \text{GeV}^2$ for photon-proton center-of-mass energies $60 < W_{\gamma^*p} < 180\ \text{GeV}$ [125]. The data sample was restricted to transverse momenta less than $0.8\ \text{GeV}^2$. They measured the ratio of the dissociative cross section to that for elastic production to be $0.63 \pm 0.11 \,(\text{stat}) \pm 0.11 \,(\text{sys})$. The t slope was found to be

$$b \;=\; 2.0 \pm 0.5 \,(\text{stat}) \pm 0.5 \,(\text{sys})\; \text{GeV}^{-2}.$$

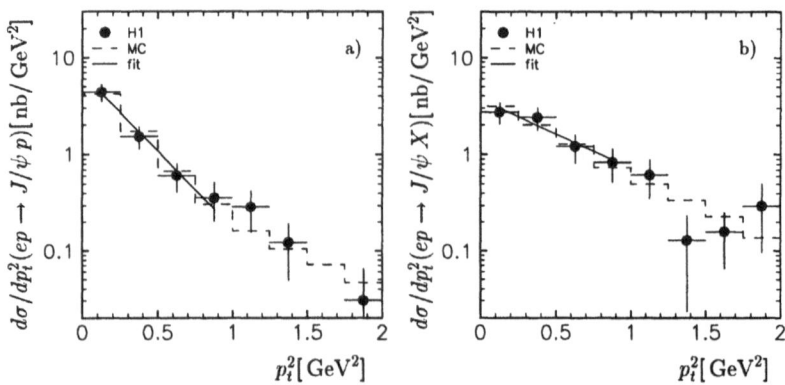

Fig. 3.28. The t dependence of the cross section for elastic J/ψ photoproduction (a) compared to that for proton dissociative J/ψ production (b) from [136]. The error bars show the systematic and statistical uncertainties added in quadrature. The solid lines are the fits assuming an exponential dependence as described in the text. The dashed lines indicate the results of simulations

This value is comparable to the value for the t slope found in J/ψ photoproduction with proton dissociation and significantly smaller than the value determined in elastic ρ^0 production at high Q^2. Such small values for the t slope ($< 4 \text{ GeV}^{-2}$) indicate that the interaction investigated has a substantial contribution from processes which resolve the proton.

A fit to the Q^2 dependence Q^{-2a} measured in diffractive ρ^0 production accompanied by proton dissociation yielded the result

$$a = 2.8 \pm 0.6 \text{ (stat)} \pm 0.3 \text{ (sys)},$$

similar to that for elastic production. A helicity analysis determined a value for the matrix element r_{00}^{04}:

$$r_{00}^{04} = 0.78 \pm 0.10 \text{ (stat)} \pm 0.05 \text{ (sys)},$$

consistent with the $\cos\theta_h$ dependence found in the case of elastic production. The latter results encourage the conclusion that the ρ^0 production process and the proton dissociative process with little momentum transferred to the proton are weakly correlated, if at all.

The ZEUS collaboration has employed two methods to identify samples of ρ^0 photoproduction events in which the proton dissociates using the data recorded in 1994 [126]. The first relied on the observation of energy depositions in forward scintillation counters (PRT) while the second was based on detection of a proton in the LPS carrying less than 98% of the beam momentum, inconsistent with an elastic scatter. Figure 3.29 shows the p_T^2 distributions corresponding to these two independent data samples, as well as the result of an exponential fit to the scintillation-counter-based sample, which determined the t slope to be

$$b = 5.3 \pm 0.3 \text{ (stat)} \pm 0.7 \text{ (sys)} \text{ GeV}^{-2}.$$

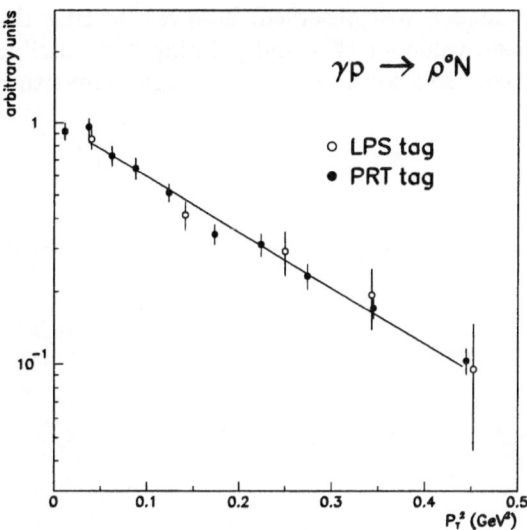

Fig. 3.29. The t dependence of the cross section for ρ^0 photoproduction accompanied by dissociation of the proton as measured by the ZEUS collaboration with data recorded in 1994 [126]. The closed circles indicate a data sample based on the observation of depositions in forward scintillation counters (PRT). The open circles represent a sample in which a proton was detected in the Leading Proton Spectrometer with momentum inconsistent with that of an elastic scatter. The line shows the results of an exponential fit to the PRT-selected data

The sample identified by the direct detection of a forward proton exhibited a slope of

$$b = 5.3 \pm 0.8 \text{ (stat)} \pm 1.1 \text{ (sys) GeV}^{-2}.$$

These slopes are each consistent with a value approximately half that observed for ρ^0 photoproduction without proton dissociation, similar to the slopes measured in J/ψ photoproduction and in ρ^0 production at high Q^2.

The ZEUS collaboration has presented a preliminary helicity analysis of ρ^0 photoproduction at high $|t|$ ($0.5 < |t| < 4.0$ GeV^2) [128]. Given such high momentum transfer at the proton vertex, the dominant contribution is expected from simulation studies to be from proton-dissociative events. The identification of the high-$|t|$ event sample was made possible by a small-angle electron tagger installed 44 m from the interaction point in the positron flight direction prior to the 1995 running period. This photon-tagging facility limited the photon virtuality to $Q^2 < 10^{-2}$ GeV^2, rendering its contribution to t negligible compared to that of the transverse momentum of the ρ^0 measured in the central detector. The polar decay angle distributions obtained indicated that the ρ^0 mesons are predominantly transversely polarized. The analysis averaged over the angle Φ_{h}, eliminating any contribution to the azimuthal angular dependence from the linear photon polarization (see (3.2.41)). A significant azimuthal angular dependence was observed, however, forcing the preliminary conclusion of a nonzero value for r_{1-1}^{04} and pointing to a possible breakdown of s-channel helicity conservation in the case of high momentum transfer at the proton vertex.

4. Future Developments

An appreciation of the richness of the future HERA research program is given by a perusal of the proceedings of the HERA workshop of 1995/96 [155]. The contents make it clear that the investigations published until now represent the tip of an iceberg. The purpose of this section is to give the reader a quantitative idea of which topics in the subfields of total photon–proton cross sections and exclusive vector-meson production will be extended, and of those which will become newly accessible.

4.1 Luminosity Schedule

Aside from a few specialized investigations which have been brought to conclusion, the data recorded during the 1995 and 1996 running periods have been under active analysis since they were obtained. Thus a fairly mature knowledge of the systematic issues associated with these data has been obtained within the ZEUS and H1 collaborations. Conclusive results with an integrated luminosity of ≈ 35 pb^{-1}, compared to those covered in this review, which are based on a luminosity of 7 pb^{-1}, are now being published.

Prior to the 1996/97 winter shutdown in HERA operation an agreement between the HERA collaborations and the DESY directorate to continue operation in two-year cycles was reached. This decision was motivated in part by the experience in past years of impressive improvement in accelerator operation during the running periods due to an increased familiarity with tuning algorithms and hardware reliability, which became obsolete during the shutdowns. The decision was also influenced by the scheduling of two extensive improvement projects scheduled to take place in the coming years: an upgrade of the electron ring vacuum system, which will allow the reintroduction of electrons with little loss in beam lifetime, and an ambitious luminosity upgrade system involving the installation of quadrupole magnets near the interaction region within the H1 and ZEUS detectors. The upgrade of the electron ring will occur during the 1997/98 shutdown and can be expected to yield an increase within a single year of more than an order of magnitude over the 2 pb^{-1} of electron–proton data presently available.

A design study for a luminosity upgrade [156] concluded that the present HERA design is limited to an integrated luminosity of 40–50 pb^{-1} per year, and proposed new optics, including low-β quadrupole magnets in the detectors, which result in an increase in instantaneous luminosity by more than a factor of four. Integrated luminosities of 100 pb^{-1} in the first year of operation after the upgrade and 1 fb^{-1} by the end of the year 2005 appear feasible according to the results of this study. The installation of quadrupole magnets near the interaction regions of the H1 and ZEUS experiments, which will take place during the 1999/2000 shutdown, entails severe consequences for the operation of small-angle detectors. Noteworthy among these is the necessary increase in the transverse momentum spread in the proton beam, which will increase by nearly a factor of two. The emittance of the beam is furthermore a function of the time since injection, and will therefore introduce severe systematic uncertainties in the measurement of t distributions via direct observation of the scattered proton in investigations of diffractive processes. This luminosity upgrade is, however, of decisive importance for the statistically-limited studies at high Q^2 and for those of the photoproduction of heavy vector mesons.

4.2 Beam Energies

The opportunity to explore various kinematic regions by varying the HERA beam energies has not yet been exploited, aside from the minor upgrade of the electron beam energy from 26.7 to 27.5 GeV in 1994. The most frequently discussed motivation for varying the beam momenta has been the systematically difficult measurement of the longitudinal proton structure function $F_{\mathrm{L}}(x,Q^2)$. Tentative plans to take data with a yet-unspecified lower proton beam energy have been made for the 1997 running period. Further impetus to this decision has been provided by the possibility of extending the energy range covered for the total photoproduction cross section measurement using the additional electron-tagging facilities installed in the ZEUS experiment prior to the 1995 running period. With the data recorded in 1995 and 1996 at the nominal proton energy of 820 GeV it is expected that the energy range $80 < W_{\gamma^*\mathrm{p}} < 300$ GeV can be covered and with appropriately lower proton energies in 1997 the range $55 < W_{\gamma^*\mathrm{p}} < 210$ GeV would become accessible, increasing sensitivity to the value of α'_{IP} in the phenomenological Pomeron trajectory.

The primary considerations providing upper bounds on the HERA beam energies are those of consequence for the reliability of HERA operation, an essential contributor to limits on the integrated luminosity over the first five years. To increase the lepton beam energy from 27.5 GeV to the design value of 30 GeV, improvements in the reliability of the superconducting RF cavities would be necessary. All proton beam superconducting dipoles have been tested at currents corresponding to a proton beam energy exceeding 1 TeV. However, no information concerning their reliability under long-term operation at such currents is available. In the absence of physics arguments which are convincing even under conditions of reduced machine reliability, no consensus on upgrading HERA to higher energies has been reached.

4.3 Detector Upgrades

The ZEUS electron tagger [128] is an example of several ZEUS and H1 detector upgrades for which data is already available and from which results can be expected during the coming year. Of note is the scintillating-fiber forward proton detector of H1, which will provide clean identification of elastic proton interactions, as the ZEUS Leading Proton Spectrometer has done. Prior to the 1998 running period the H1 collaboration will install a small calorimeter in the rear direction to perform measurements in a Q^2 range comparable to that covered by the ZEUS BPC.

4.4 Beam Polarization

At present it is planned to install electron spin rotators in the electron beam line during the 1997/98 shutdown, permitting the longitudinal polarization of the lepton beams. Extensive studies of polarization algorithms during HERA operation since 1993 have resulted in the routine achievement of transverse polarizations in excess of 50%. While transverse polarization has negligible influence on the decay angular distributions used for the helicity analyses in vector-meson production, longitudinal polarization plays an important role in that it represents the only way to distinguish the components of the spin density matrix corresponding to circular photon polarization [153].

A discussion of the possibility of running polarized proton beams at HERA can be found in [157], but no definitive decision concerning such a project has been made.

4.5 Nuclear Beams

Simple model-independent considerations easily show why there has been such great theoretical interest in the introduction of nuclear beams at HERA. (For a compilation, see [158].) Our calculations of virtual photon lifetime in Sect. 3.1 showed that the bulk of the HERA data cover a kinematic region where the photon lifetime exceeds the interaction time associated with the size of the proton in its rest frame. The interaction time increases with the extent of the hadronic target and so for nuclear beams the threshold region moves toward the kinematic region where HERA provides greater statistical power. The result is that the nuclear target size represents a further parameter which can be tuned to scan the space–time structure of deep inelastic scattering, along with the photon virtuality and the mass of diffractively-produced vector mesons. Striking effects are predicted in the Q^2 dependence of diffractive production of radially-excited vector mesons off nuclei [159]. Further arguments are based on the precepts of QCD, which result in the expectation of strong nuclear enhancements of nonlinear effects at low x [89, 91, 160]. Extensive treatment of the topic of the electroproduction of vector mesons off nuclei may be found in the review of Bauer et al. [49], who consider such investigations important as tests of vector-dominance models.

The investigations by the Light and Heavy Nuclei Working Group in the workshop *Future Physics at HERA* concluded that substantial investment in the development of an ion source would be necessary to introduce nuclear beams. Despite strong interest in a deuteron beam from other working groups and the apparent feasibility of nuclear beams of nuclear weights as high as that of sulfur, this project remains only at the discussion stage.

5. Summary

Results obtained by the ZEUS and H1 collaborations from investigations of high-energy electron-proton interactions during the first five years of operation of the HERA collider have demonstrated the versatility of this experimental technique. These studies have yielded quantititative insight into questions of proton structure, electroweak cross sections at the unification scale, the hypothesized existence of particles and interactions unforeseen in the Standard Model, diffractive processes at high energy and the hadronic interactions of photons both virtual and real. This review began with a historical perspective on the decision to build an electron–proton collider, and described the operation of the HERA accelerator complex during its first years of operation. Following a description of the experimental apparatus employed by the ZEUS and H1 collaborations, a general overview of the physics topics addressed in the investigations of photon–proton interactions at HERA was given. Our introduction concluded by placing the recent studies of diffractive vector-meson production at HERA in the rich historical context of past and present experimental investigations of vector-meson lepto- and photoproduction.

The recently obtained wealth of new information has motivated a remarkable variety of new theoretical calculations. These span a wide range of physical concepts, requiring tools and ideas developed separately for the description of soft diffractive hadronic processes and for that of hard processes subject to the application of perturbative calculational techniques. In Chap. 2 we presented a discussion of general theoretical considerations followed by synopses of topical models in this field.

Chapter 3 addressed the studies at HERA of the interactions of photons with protons in a range of photon virtuality Q^2 between 10^{-10} and 10^3 GeV^2, and of the Bjorken scaling variable x between 10^{-5} and 10^{-1}. After showing that in the low-x kinematic region covered by the HERA data the virtual photon lifetime far exceeds the interaction time, we defined the total virtual-photon–proton cross section and presented measurements of it for energies of the photon in the proton rest frame between 400 and 40 000 GeV. We concluded that a region of transition from an energy dependence characteristic of phenomenological Regge-theory-based models to that described by QCD evolution lies in the region of photon virtuality between 0.1 and 3 GeV^2.

These preliminary studies will be extended in order to further investigate the underlying dynamics governing the transition region.

This brief summary of issues concerning the total inclusive virtual-photon–proton cross section served to introduce our main topic, the exclusive production of single neutral vector mesons in photon–proton interactions at high energy. A comprehensive discussion of results published by the H1 and ZEUS collaborations prior to January 1, 1997 was presented. The experimental issues of technique, resolution, efficiencies, and backgrounds were covered for the measurements of ρ^0, ω, ϕ, and J/ψ mesons, and for the initial observations of ρ' and $\psi(2S)$ mesons. The energy dependence of the elastic cross sections at high photon virtuality was shown to be steep, similar to that observed for the inclusive cross section. In photoproduction the energy dependence was found to be much weaker, comparable to that for the total photoproduction cross section, excepting that for J/ψ production. These observations are consistent with calculations performed in the context of perturbative QCD, in which the necessary hard scale is given either by the photon virtuality or by the vector-meson mass. We continued our description of results by considering the Q^2 dependence of the cross section, which was seen to be consistent with a dependence of the form $(Q^2 + M_V^2)^{-a}$, with $2 < a < 3$, but for which the experimental uncertainties are at a level which does not clearly distinguish the exponent for the various vector mesons. There are, however, significant indications that the Q^2 dependence is weaker than Q^{-6}, in accordance with expectations based on perturbative QCD calculations. This comparison is mitigated by the ambiguity introduced because of the unseparated contributions to the measured cross section from transversely and longitudinally polarized photons.

The exponential slopes of the differential cross sections as functions of the momentum transfer at the proton vertex indicated that the interaction sizes for the ϕ and J/ψ mesons are distinctly smaller than those for the ρ^0 and ω mesons in photoproduction. There are also preliminary indications that the interaction size for elastic ρ^0 production decreases with Q^2. The measurements cover a Q^2 range in which the exponential slope approaches a value consistent with that expected from the proton alone.

Comparisons of the production rates for ϕ and ρ^0 mesons for photon virtuality exceeding 5 GeV^2 have shown that the ratio is consistent with the value of 2/9 expected from a simple quark-counting rule and a flavor-independent production mechanism, contrasting with the relative suppression of approximately a factor of three in ϕ photoproduction. Even more impressive are the results for J/ψ production at high Q^2, which show that the suppression by two orders of magnitude relative to the ρ^0 as measured in photoproduction is reduced to a mere factor of about three.

The helicity analyses of the vector-meson decay angle distributions were considered in detail, culminating in a preliminary determination of the ratio of longitudinal to transverse photon cross sections as a function of Q^2

in exclusive ρ^0 production under the assumption of s-channel helicity conservation. This ratio rises from a value consistent with zero at the level of a few percent in photoproduction to a value exceeding unity for Q^2 greater than 5 GeV^2. This Q^2 dependence is particularly sensitive to the mixture of point-like and nonperturbative processes in the vector-meson production mechanism.

Our consideration of experimental results concluded with a brief description of measurements of vector-meson production in processes where dissociation of the proton is observed. Such data sets have been identified in ρ^0 and J/ψ photoproduction and in ρ^0 production for $Q^2 > 7$ GeV^2. A much weaker dependence on the momentum transfer at the proton vertex $|t|$ characterizes these processes, with the t slope measured to be only about half that of the purely elastic interaction. Helicity analyses applied to the high-Q^2 ρ^0 production at low $|t|$ showed characteristics similar to those observed for event samples without proton dissociation, indicating that the ρ^0 production mechanism and the dissociation of the proton proceed independently in this low $|t|$ region. However, a preliminary helicity analysis applied to an event sample which was restricted to high $|t|$ ($|t| > 0.5$ GeV^2) showed evidence for the breakdown of s-channel helicity conservation.

The H1 and ZEUS collaborations are presently completing analyses of data sets obtained from a combined integrated luminosity approximately a factor of five greater than that covered by the analyses described in this review. Furthermore, a luminosity upgrade, as well as upgraded apparatus and polarization of electron and positron beams, is planned for the coming years. The introduction of nuclear beams is under consideration.

6. Concluding Remarks

The introduction of nonabelian field theory in the 1970s as a description of the strong interaction has proven immensely successful at transverse momentum scales exceeding the confinement scale. Soft hadronic processes are well described by phenomenological fits inspired by Regge theory. But there has been little progress in developing a theoretical picture which offers a unified understanding of the relationship between these two distinct dynamical descriptions. The early investigations of high-energy electron–proton interactions at HERA described in this review yield new information on the transition region between the hard and soft hadronic interactions of the photon. The high virtual photon energies at HERA provide a clean environment for the study of diffractive processes. An example of such interactions is the elastic production of vector mesons, which has been shown to provide a further handle on the transition between hard and soft processes, as the photoproduction of J/ψ mesons exhibits an energy dependence characteristic of hard processes, while that of the lighter vector mesons has been shown to obey scaling laws typical of soft interactions. Variation of photon virtuality and vector-meson mass has been used to scan the space–time structure of the photon–proton interaction, yielding information on the wave function of the virtual photon and on that of the vector meson. Measurements of vector-meson production ratios have shown that flavor asymmetries decrease with increasing photon virtuality. Such revelations, together with the future schedule of HERA, justify confidence in the potential of these investigations to further contribute to our understanding of the strong interaction.

References

1. *Proceedings of the Seminar on $e^- p$ and $e^- e^+$ Storage Rings*, edited by J.K. Bienlein, I. Dammann, and H. Wiedemann (DESY, Hamburg, Germany, 1973);
   H. Gerke et al., DESY Report **H–72–22** (1972) .
2. S. Conetti et al., in *Proceedings of the 11th International Conference on High Energy Accelerators*, edited by W.S. Newman (Birkhäuser, Geneva, Switzerland, 1980), p. 189.
3. *Proceedings of the Study of an ep-Facility for Europe*, edited by U. Amaldi (DESY, Hamburg, Germany, 1979).
4. DESY Report **H–81/10** (1981) ;
   B.H. Wiik, in *Proceedings of the 11th International Conference on High Energy Accelerators*, edited by W.S. Newman (Birkhäuser, Geneva, Switzerland, 1980), p. 120;
   G.-A. Voss, in *Proceedings of the 12th International Conference on High Energy Accelerators*, edited by F.T. Cole and R. Donaldson (FNAL, Batavia, Illinois, 1983), p. 29.
5. G.-A. Voss and B.H. Wiik, Ann. Rev. Nucl. Part. Sci. **44** (1994) 413;
   F. Willeke, Int. J. Mod. Phys. **Proc. Suppl. 2A** (1993) 28;
   R. Brinkmann, DESY Report **M–95–08A** (1995) .
6. The H1 Collaboration, DESY Report **96–001** (1996) .
7. J. Buerger et al., Nucl. Instrum. Methods **A279** (1989) 217.
8. The H1 Calorimeter Group, Nucl. Instrum. Methods **A336** (1993) 460.
9. The ZEUS Collaboration, *The ZEUS Detector*, Status Report, 1993.
10. R. Klanner, Nucl. Instrum. Methods **A 265** (1988) 200;
    U. Behrens et al., Nucl. Instrum. Methods **A 289** (1990) 115;
    A. Andresen et al., Nucl. Instrum. Methods **A 290** (1990) 95;
    A. Bernstein et al., Nucl. Instrum. Methods **A 336** (1993) 23.
11. J.E. Brau and T.E. Gabriel, Nucl. Instrum. Methods **A 238** (1985) 489;
    R. Wigmans, Nucl. Instrum. Methods **A 259** (1987) 389;
    H. Brückmann et al., Nucl. Instrum. Methods **A 263** (1988) 136.
12. M. Derrick et al., Nucl. Instrum. Methods **A 309** (1991) 77;
    A. Andresen et al., Nucl. Instrum. Methods **A 309** (1991) 101.
13. J.A. Crittenden, in *Proceedings of the Fifth International Conference on Calorimetry in High Energy Physics, Brookhaven National Laboratory*, edited by H.A. Gordon and D. Rueger (World Scientific, Singapore, 1994), p. 58.
14. N. Harnew et al., Nucl. Instrum. Methods **A 279** (1989) 290;
    C.B. Brooks et al., Nucl. Instrum. Methods **A 283** (1989) 473;
    B. Foster et al., Nucl. Instrum. Methods **A 338** (1993) 181.
15. A. Bamberger et al., Nucl. Instrum. Methods **A 382** (1996) 419.
16. G. Abbiendi et al., Nucl. Instrum. Methods **A 333** (1993) 342.
17. A. Caldwell et al., Nucl. Instrum. Methods **A 321** (1992) 356;

W. Buttler et al., Nucl. Instrum. Methods **A 277** (1989) 217;
A. Caldwell et al., DESY Report **92–130** (1992) ;
L. Hervas, *The Readout for the ZEUS Calorimeter*, Ph.D. thesis, Universidad Autónoma de Madrid, 1991, DESY F35D–91–01;
W. Sippach et al., IEEE Trans. **NS 36** (1989) 465.

18. W.H. Smith et al., Nucl. Instrum. Methods **A 355** (1995) 278.

19. The ZEUS Collaboration, Z. Phys. **C 73** (1997) 253.

20. R. Sacchi, *Studio di eventi diffrattivi con uno spettrometro per protoni a ZEUS*, Ph.D. thesis, University of Torino, 1996.

21. M. Breidenbach et al., Phys. Rev. Lett. **23** (1969) 935.

22. J.D. Bjorken, Phys. Rev. **179** (1969) 1547;
R.P. Feynman, *Photon–Hadron Interactions* (W.A. Benjamin, Inc., 1972);
R.P. Feynman, Phys. Rev. Lett. **23** (1969) 1415;
J.B. Kogut and D.E. Soper, Phys. Rev. **D 1** (1970) 2901;
S.D. Drell, D.J. Levy, and T.M. Yan, Phys. Rev. Lett. **22** (1969) 744.

23. J.D. Bjorken, J.B. Kogut, and D.E. Soper, Phys. Rev. **D 3** (1971) 1382.

24. E. Byckling and K. Kajantie, *Particle Kinematics* (Wiley, New York, 1973).

25. G. Altarelli and G. Martinelli, Phys. Lett. **B 76** (1978) 89.

26. The H1 Collaboration, Phys. Lett. **B 299** (1993) 385;
The H1 Collaboration, Nucl. Phys. **B 407** (1993) 515;
The H1 Collaboration, Phys. Lett. **B 321** (1994) 161;
The ZEUS Collaboration, Phys. Lett. **B 303** (1993) 183.

27. The H1 Collaboration, Nucl. Phys. **B 439** (1995) 471;
The ZEUS Collaboration, Phys. Lett. **B 316** (1993) 412.

28. The ZEUS Collaboration, Z. Phys. **C 65** (1995) 379.

29. The ZEUS Collaboration, Z. Phys. **C 72** (1996) 399.

30. The NMC Collaboration, Phys. Lett. **B 364** (1995) 107.

31. K. Prytz, Phys. Lett. **332** (1994) 393.

32. V.N. Gribov and L.N. Lipatov, Sov. J. Nucl. Phys. **15** (1972) 438;
V.N. Gribov and L.N. Lipatov, Sov. J. Nucl. Phys. **15** (1972) 675;
L.N. Lipatov, Sov. J. Nucl. Phys. **20** (1975) 95;
Yu.L Dokshitzer, Sov. Phys. JETP **46** (1977) 641;
G. Altarelli and G. Parisi, Phys. Lett. **126** (1977) 298.

33. A. De Rújula et al., Phys. Rev. **D 10** (1974) 1649.

34. C. López and F.J. Ynduráin, Phys. Rev. Lett. **44** (1980) 1118;
F. Barreiro, C. López, and F.J. Ynduráin, Z. Phys. **C 72** (1996) 561.

35. The H1 Collaboration, Nucl. Phys. **B 470** (1996) 3.

36. The H1 Collaboration, Phys. Lett. **B 354** (1995) 494;
The H1 Collaboration, Nucl. Phys. **B 449** (1995) 3;
The ZEUS Collaboration, Phys. Lett. **B 345** (1995) 576.

37. The H1 Collaboration, Phys. Lett. **B 393** (1997) 452.

38. The H1 Collaboration, Phys. Lett. **B 298** (1993) 469;
The H1 Collaboration, Z. Phys. **C 61** (1994) 59;
The H1 Collaboration, Z. Phys. **C 63** (1994) 377;
The H1 Collaboration, Phys. Lett. **B 346** (1995) 415;
The H1 Collaboration, Nucl. Phys. **B 445** (1995) 3;
The H1 Collaboration, Phys. Lett. **B 356** (1995) 118;
The H1 Collaboration, Phys. Lett. **B 358** (1995) 412;
The H1 Collaboration, Z. Phys. **C 72** (1996) 573;
The H1 Collaboration, Nucl. Phys. **B 485** (1997) 3;
The ZEUS Collaboration, Phys. Lett. **B 306** (1993) 158;
The ZEUS Collaboration, Phys. Lett. **C 59** (1993) 231;
The ZEUS Collaboration, Z. Phys. **C 67** (1995) 81;

The ZEUS Collaboration, Z. Phys. **C 67** (1995) 93;
The ZEUS Collaboration, Phys. Lett. **B 363** (1995) 201;
The H1 Collaboration, Nucl. Phys. **B 480** (1996) 3;
The ZEUS Collaboration, Z. Phys. **C 68** (1995) 29;
The H1 Collaboration, Z. Phys. **C 72** (1996) 593;
The ZEUS Collaboration, Phys. Lett. **B 338** (1994) 483.
39. The ZEUS Collaboration, Z. Phys. **C 70** (1996) 1.
40. The H1 Collaboration, Nucl. Phys. **B 429** (1994) 477;
The H1 Collaboration, Phys. Lett. **B 348** (1995) 681;
The H1 Collaboration, Z. Phys. **C 70** (1996) 609;
The ZEUS Collaboration, Z. Phys. **C 68** (1995) 569;
The ZEUS Collaboration, Z. Phys. **C 70** (1996) 391;
The ZEUS Collaboration, Phys. Lett. **B 315** (1993) 481.
41. The ZEUS Collaboration, Phys. Lett. **B 384** (1996) 388.
42. B.D. Burow et al., DESY Report **94–215** (1994) .
43. G. Schuler and T. Sjöstrand, Nucl. Phys. **B 407** (1993) 539.
44. M. Erdmann, *The Partonic Structure of the Photon,* Springer Tracts in Modern Physics, Volume 138 (Springer, Berlin Heidelberg, 1997).
45. The H1 Collaboration, Phys. Lett. **B 297** (1992) 205;
The H1 Collaboration, Phys. Lett. **B 314** (1993) 436;
The H1 Collaboration, Phys. Lett. **B 328** (1994) 176;
The H1 Collaboration, Nucl. Phys. **B 445** (1995) 195;
The H1 Collaboration, Z. Phys. **C 70** (1996) 17;
The ZEUS Collaboration, Phys. Lett. **B 297** (1992) 404;
The ZEUS Collaboration, Phys. Lett. **B 322** (1994) 287;
The ZEUS Collaboration, Phys. Lett. **B 342** (1995) 417;
The ZEUS Collaboration, Phys. Lett. **B 354** (1995) 163;
The ZEUS Collaboration, Phys. Lett. **B 384** (1995) 401;
The ZEUS Collaboration, Phys. Lett. **B 348** (1995) 665.
46. The H1 Collaboration, Nucl. Phys. **B 435** (1995) 3;
The ZEUS Collaboration, Z. Phys. **C 67** (1995) 227.
47. The ZEUS Collaboration, Phys. Lett. **B 346** (1995) 399;
The ZEUS Collaboration, Phys. Lett. **B 356** (1995) 129;
The ZEUS Collaboration, Phys. Lett. **B 369** (1995) 55.
48. The H1 Collaboration, Nucl. Phys. **B 472** (1996) 32;
The ZEUS Collaboration, Phys. Lett. **B 349** (1995) 225.
49. T.H. Bauer et al., Rev. Mod. Phys. **50** (1978) 261.
50. J. Ballam et al., Phys. Rev. **D 5** (1972) 545.
51. P. Joos et al., Nucl. Phys. **B 113** (1976) 53.
52. V.L. Auslander et al., Sov. J. Nucl. Phys. **9** (1969) 69;
K.C. Moffeit et al., Nucl. Phys. **B 29** (1971) 349;
D. Benaksas et al., Phys. Lett. **39** (1972) 289;
L.M. Barkov et al., Nucl. Phys. **B 256** (1985) 365.
53. R.M. Egloff et al., Phys. Rev. Lett. **43** (1979) 657;
R.M. Egloff et al., Phys. Rev. Lett. **43** (1979) 1545;
D. Aston et al., Nucl. Phys. **B 209** (1982) 56.
54. J. Park et al., Nucl. Phys. **B 36** (1972) 404;
J.T. Dakin et al., Phys. Rev. **D 8** (1973) 687;
J. Ballam et al., Phys. Rev. **D 10** (1974) 765;
C. del Papa et al., Phys. Rev. **D 19** (1979) 1303;
D.G. Cassel et al., Phys. Rev. **D 24** (1981) 2787;
I. Cohen et al., Phys. Rev. **D 25** (1982) 634.
55. W.R. Francis et al., Phys. Rev. Lett. **38** (1977) 633;

W.D. Shambroom et al., Phys. Rev. **D 26** (1982) 1.

56. The EMC Collaboration, Phys. Lett. **B 161** (1985) 203.
57. The EMC Collaboration, Z. Phys. **C 39** (1988) 169.
58. The NMC Collaboration, Z. Phys. **C 54** (1992) 239.
59. The NMC Collaboration, Nucl. Phys. **B 429** (1994) 503.
60. The E665 Collaboration, Z. Phys. **C 72** (1997) 237;
    The E665 Collaboration, PA03–009, XXVIII International Conference on High Energy Physics, Warsaw, Poland, July 25–31, 1996.
61. G.T. Jones et al., Z. Phys. **C 58** (1993) 375;
    B. Kopeliovich and P. Marage, Int. J. Mod. Phys. **A 8** (1993) 513.
62. B. Gittelman et al., Phys. Rev. Lett. **35** (1975) 1616.
63. B. Knapp et al., Phys. Rev. Lett. **34** (1975) 1040;
    T. Nash et al., Phys. Rev. Lett. **36** (1976) 1233.
64. U. Camerini et al., Phys. Rev. Lett. **35** (1975) 483.
65. S.D. Holmes, W. Lee, and J.E. Wiss, Ann. Rev. Nucl. Part. Sci. **35** (1985) 397.
66. M. Binkley et al., Phys. Rev. Lett. **48** (1982) 73;
    B.H. Denby et al., Phys. Rev. Lett. **52** (1984) 795.
67. A.R. Clark et al., Phys. Rev. Lett. **43** (1979) 187.
68. The EMC Collaboration, Nucl. Phys. **B 213** (1983) 1.
69. The NMC Collaboration, Phys. Lett. **B 332** (1994) 195.
70. R. Barate et al., Z. Phys. **C 33** (1987) 505.
71. P.L. Frabetti et al., Phys. Lett. **B 316** (1993) 197.
72. H. Abramowicz, L.L. Frankfurt, and M. Strikman, DESY Report **95–047** (1995) , HEP–PH/95-03-437, published in SLAC Summer Inst., 1994.
73. M.L. Perl, *High Energy Hadron Physics* (Wiley, New York, 1974).
74. P.D.B. Collins, *An Introduction to Regge Theory and High Energy Physics* (Cambridge University Press, 1977).
75. J.C. Collins et al., LANL Preprint HEP–PH/96-11-433 (1996).
76. A. Donnachie and P.V. Landshoff, Phys. Lett. **B 348** (1995) 213.
77. A. Donnachie and P.V. Landshoff, Phys. Lett. **B 296** (1992) 227;
    R.M. Barnett et al., Rev. Mod. Phys. **68** (1996) 611.
78. H. Fraas and D. Schildknecht, Nucl. Phys. **B 14** (1969) 543.
79. J.J. Sakurai, Phys. Rev. Lett. **22** (1969) 981;
    J.J. Sakurai, *Currents and Mesons* (University of Chicago Press, Chicago, IL, 1969).
80. L.P.A. Haakman et al., LANL Preprint HEP–PH/95-07-394 (1995).
81. L.L. Jenkovszky et al., LANL Preprint HEP–PH/96-08-384 (1996).
82. M.A. Pichowsky and T.S.H. Lee, Phys. Lett. **379** (1996) 1.
83. H.G. Dosch et al., LANL Preprint HEP–PH/96-08-203 (1996).
84. G. Niesler et al., Phys. Lett. **389** (1996) 157.
85. The ZEUS Collaboration, Z. Phys. **C 69** (1995) 39.
86. M.G. Ryskin, Z. Phys. **C 57** (1993) 89.
87. M. Arneodo, L. Lamberti, and M. Ryskin, Comp. Phys. Commun. **100** (1996) 195.
88. M.G. Ryskin et al., LANL Preprint HEP–PH/95-11-228 (1995).
89. S.J. Brodsky et al., Phys. Rev. **D 50** (1994) 3134.
90. S.J. Brodsky and G.P. Lepage, Phys. Rev. **D 22** (1980) 2157.
91. L. Frankfurt et al., Phys. Rev. **D 54** (1996) 3194.
92. P. Hoodbhoy, LANL Preprint HEP–PH/96-11-207 (1996).
93. A. Martin et al., LANL Preprint HEP–PH/96-09-448 (1996).
94. J. Nemchik et al., Phys. Lett. **374** (1996) 199;
    J. Nemchik et al., Phys. Lett. **341** (1994) 228.
95. J. Nemchik et al., LANL Preprint HEP–PH/96-05-231 (1996).

96. L.N. Lipatov, LANL Preprint HEP–PH/96-10-276 (1996);
    L.N. Lipatov, Sov. J. Nucl. Phys. **23** (1976) 642;
    V.S. Fadin et al., Phys. Lett. **B 60** (1975) 50;
    E.A. Kuraev et al., Sov. Phys. JETP **44** (1976) 443;
    E.A. Kuraev et al., Sov. Phys. JETP **45** (1977) 199;
    Ya.Ya. Balitsky and L.N. Lipatov, Sov. J. Nucl. Phys. **28** (1978) 822.
97. J. Amundson et al., LANL Preprint HEP–PH/96-01-298 (1996).
98. G.T. Bodwin et al., Phys. Rev. **D 51** (1995) 1125.
99. D.Yu. Ivanov, Phys. Rev. **D 53** (1996) 3564;
    I.F. Ginzburg and D.Yu. Ivanov, Phys. Rev. **D 54** (1996) 5523.
100. J. Bartels et al., Phys. Lett. **B 375** (1996) 301.
101. J.R. Forshaw and M.G. Ryskin, Z. Phys. **C 68** (1995) 137.
102. The H1 Collaboration, Phys. Lett. **B 299** (1993) 374.
103. The ZEUS Collaboration, Phys. Lett. **B 293** (1992) 465.
104. The H1 Collaboration, Z. Phys. **C 69** (1995) 27.
105. The ZEUS Collaboration, Z. Phys. **C 63** (1994) 391.
106. B.L. Ioffe, Phys. Lett. **B 30** (1969) 123.
107. B.L. Ioffe, V.A. Khoze, and L.N. Lipatov, *Hard Processes* (North Holland, Amsterdam, 1984).
108. L. Hand, Phys. Rev. **120** (1963) 1834.
109. F. Gilman, Phys. Rev. **167** (1968) 1365.
110. The ZEUS Collaboration, Z. Phys. **C 69** (1996) 607.
111. The BCDMS Collaboration, Phys. Lett. **B 223** (1989) 485.
112. The E665 Collaboration, Phys. Rev. **D 54** (1996) 3006.
113. D.O. Caldwell et al., Phys. Rev. Lett. **40** (1978) 1222;
    S.I. Alekhin et al., CERN–HERA **87-001** (1987) .
114. A. Donnachie and P. Landshoff, Z. Phys. **C 61** (1994) 139.
115. H. Abramowicz et al., Phys. Lett. **B 269** (1991) 465;
    A. Marcus, *Energy Dependence of the γ^*p Cross Section*, Master's thesis, Tel-Aviv University, 1996.
116. The ZEUS Collaboration, PA02–025, XXVIII International Conference on High Energy Physics, Warsaw, Poland, July 25–31, 1996.
117. M. Glück, E. Reya, and A. Vogt, Z. Phys. **C 48** (1990) 471;
    M. Glück, E. Reya, and A. Vogt, Z. Phys. **C 53** (1992) 127;
    M. Glück, E. Reya, and A. Vogt, Z. Phys. **C 67** (1995) 433.
118. A. Levy, LANL Preprint HEP–PH/96-08-9 (1996).
119. The ZEUS Collaboration, PA02–028, XXVIII International Conference on High Energy Physics, Warsaw, Poland, July 25–31, 1996.
120. The H1 Collaboration, PA02–085, XXVIII International Conference on High Energy Physics, Warsaw, Poland, July 25–31, 1996.
121. N. Cartiglia, LANL Preprint HEP–PH/97-03-245 (1997), published in SLAC Summer Inst., 1996.
122. The H1 Collaboration, Nucl. Phys. **B 463** (1996) 3.
123. The ZEUS Collaboration, Phys. Lett. **B 356** (1995) 601.
124. The H1 Collaboration, Nucl. Phys. **B 468** (1996) 3.
125. The H1 Collaboration, PA02–065, XXVIII International Conference on High Energy Physics, Warsaw, Poland, July 25–31, 1996.
126. The ZEUS Collaboration, PA02–050, XXVIII International Conference on High Energy Physics, Warsaw, Poland, July 25–31, 1996.
127. The ZEUS Collaboration, PA02–053, XXVIII International Conference on High Energy Physics, Warsaw, Poland, July 25–31, 1996.
128. The ZEUS Collaboration, PA02–051, XXVIII International Conference on High Energy Physics, Warsaw, Poland, July 25–31, 1996.

129. The H1 Collaboration, PA01–088, XXVIII International Conference on High Energy Physics, Warsaw, Poland, July 25–31, 1996.
130. The ZEUS Collaboration, Z. Phys. **C 73** (1996) 73.
131. The ZEUS Collaboration, Phys. Lett. **B 377** (1996) 259.
132. The ZEUS Collaboration, Phys. Lett. **B 380** (1996) 220.
133. The H1 Collaboration, PA02–064, XXVIII International Conference on High Energy Physics, Warsaw, Poland, July 25–31, 1996.
134. The ZEUS Collaboration, Phys. Lett. **B 350** (1995) 120.
135. The ZEUS Collaboration, PA02–047, XXVIII International Conference on High Energy Physics, Warsaw, Poland, July 25–31, 1996.
136. The H1 Collaboration, Nucl. Phys. **B 472** (1996) 3.
137. The H1 Collaboration, PA02–086, XXVIII International Conference on High Energy Physics, Warsaw, Poland, July 25–31, 1996.
138. N.N. Achasov and V.A. Karnekov, Sov. J. Nucl. Phys. **38** (1983) 736.
139. S. Bentvelsen, J. Engelen, and P. Kooijman, in *Proceedings of the Workshop on Physics at HERA*, edited by W. Buchmüller and G. Ingleman (DESY, Hamburg, Germany, 1992), p. 23.
140. K. Piotrzkowski, *Experimental Aspects of the Luminosity Measurement in the ZEUS Experiment*, Ph.D. thesis, University of Hamburg, 1993, DESY F35D-93-06.
141. P. Söding, Phys. Lett. **19** (1966) 702.
142. J.D. Jackson, Nuovo Cimento **34** (1964) 1644.
143. S. Drell, Phys. Rev. Lett. **5** (1960) 278.
144. M. Ross and L. Stodolsky, Phys. Rev. **149** (1966) 1172;
     R. Spital and D.R. Yennie, Phys. Rev. **D 9** (1974) 126;
     J. Pumplin, Phys. Rev. **D 2** (1970) 1859;
     T.H. Bauer, Phys. Rev. Lett. **25** (1970) 485;
     T.H. Bauer, Phys. Rev. Lett. **25** (1970) 704, (E);
     T.H. Bauer, Phys. Rev. **D 3** (1971) 2671.
145. The H1 Collaboration, DESY Report **96–162** (1996) .
146. The H1 Collaboration, PA03–048, XXVIII International Conference on High Energy Physics, Warsaw, Poland, July 25–31, 1996.
147. The H1 Collaboration, Phys. Lett. **B 338** (1994) 507.
148. P.G.O. Freund, Nuovo Cimento **A 48** (1967) 541.
149. R. Erbe et al., Phys. Rev. **175** (1968) 1669.
150. J.R. Cudell, Nucl. Phys. **B 336** (1990) 1.
151. F. Gilman et al., Phys. Lett. **B 31** (1970) 387.
152. B. Humbert and A.C.D. Wright, Phys. Lett. **B 65** (1976) 463;
     B. Humbert and A.C.D. Wright, Phys. Rev. **D 15** (1977) 2503.
153. K. Schilling and G. Wolf, Nucl. Phys. **B 61** (1973) 381.
154. G. Wolf and P. Söding, in Electromagnetic Interactions of Hadrons, edited by A. Donnachie and G. Shaw, (Plenum, New York, 1978).
155. *Proceedings of the Workshop on Future Physics at HERA*, edited by G. Ingelman, A. De Roeck, and R. Klanner (DESY, Hamburg, Germany, 1996).
156. W. Bartel et al., in *Proceedings of the Workshop on Future Physics at HERA*, edited by G. Ingelman, A. De Roeck, and R. Klanner (DESY, Hamburg, Germany, 1996), p. 1095.
157. D.P. Barber et al., in *Proceedings of the Workshop on Future Physics at HERA*, edited by G. Ingelman, A. De Roeck, and R. Klanner (DESY, Hamburg, Germany, 1996), p. 1205.
158. M. Arneodo et al., in *Proceedings of the Workshop on Future Physics at HERA*, edited by G. Ingelman, A. De Roeck, and R. Klanner (DESY, Hamburg, Germany, 1996), p. 887.

159. J. Nemchik et al., LANL Preprint HEP–PH/96-05-208 (1996).
160. B.Z. Kopeliovich et al., Phys. Lett. **B 309** (1993) 179;
      L.L. Frankfurt and M.I. Strikman, Phys. Lett. **B 382** (1996) 6;
      B.Z. Kopeliovich et al., Phys. Lett. **B 324** (1994) 469;
      J. Hüfner et al., Phys. Lett. **B 383** (1996) 362.

Index

Springer Tracts in Modern Physics